W9-ATT-899

A SHEARWATER BOOK

IN SEARCH OF NATURE

EDWARD O. WILSON

IN SEARCH

OF NATURE

Illustrations by Laura Simonds Southworth

ISLAND PRESS / Shearwater Books
Washington, D.C. • Covelo, California

Library of Congress Cataloging-in-Publication Data

Wilson, Edward Osborne, 1929-
In search of nature / Edward O. Wilson.
p. cm.
"A Shearwater book."
Includes bibliographical references and index.
ISBN 1-55963-215-1 (cloth). — ISBN 1-55963-216-X (paper)
1. Philosophy of nature. 2. Man. 3. Human ecology—Philosophy
I. Title
BD581.W476 1996
113—dc20 96-11226
CIP

Printed on recycled, acid-free paper

Manufactured in the United States of America

10 9 8 7 6 5 4 3

CONTENTS

Preface *ix*

Animal Nature, Human Nature

The Serpent *3*
In Praise of Sharks *31*
In the Company of Ants *45*
Ants and Cooperation *61*

The Patterns of Nature

Altruism and Aggression *73*
Humanity Seen from a Distance *95*
Culture as a Biological Product *105*
The Bird of Paradise: The Hunter and the Poet *127*

Nature's Abundance

The Little Things That Run the World *139*
Systematics Ascending *147*
Biophilia and the Environmental Ethic *153*
Is Humanity Suicidal? *181*

Acknowledgment of Sources *201*
Index *203*

PREFACE

THE COLLECTED ESSAYS offered here, first published from 1975 through 1993, address two archetypal and hence elusive conceptions. The first is nature, that part of the world we think of as eternal, beyond us, having no need of us, and yet is the cradle of our species. The second is human nature, our essence, the way we were in the beginning, comprising those sensory and emotional capacities that join humanity into one species as surely as language and ethnic custom divide us into tribes.

The central theme of the essays is that wild nature and human nature are closely interwoven. I argue that the only way to make complete sense of either is by examining both closely and together as products of evolution. Natural history then gains more meaning, while the diversity of life, which we are so recklessly diminishing through species extinction, attains higher value. Human behavior is seen as the product not just of recorded history, ten thousand years recent, but of deep history, the combined genetic and cultural changes that created humanity over hundreds of years. We need this longer view, I believe, not only to understand our species but more firmly to secure its future.

LEXINGTON, MASSACHUSETTS

ANIMAL NATURE,

HUMAN NATURE

THE SERPENT

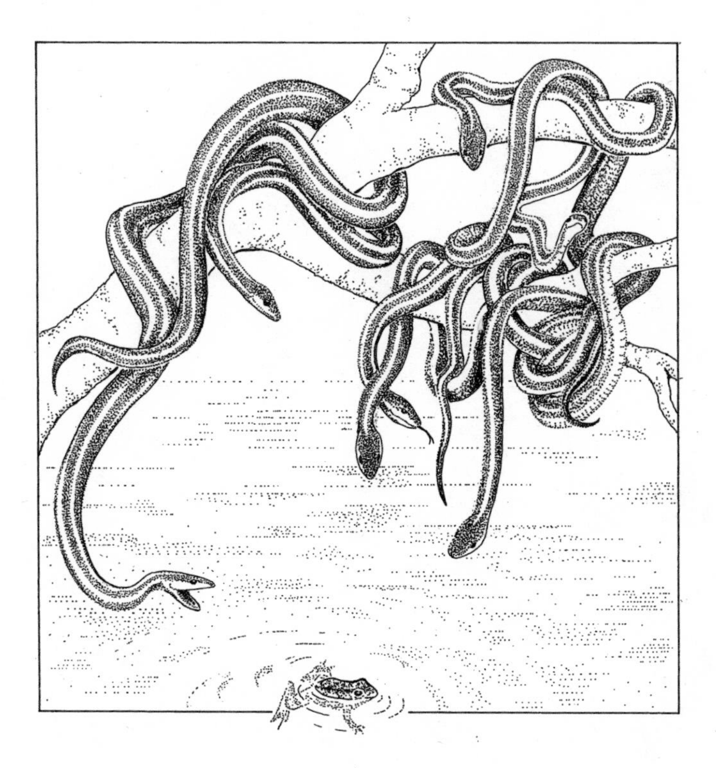

SCIENCE AND THE HUMANities, biology and culture, are bridged in a dramatic manner by the phenomenon of the serpent. Fabricated from symbols and bearing portents of magic, the snake's image enters the conscious and unconscious mind with ease during reverie and dreams. It appears without warning and departs abruptly, leaving behind not a specific memory of any real snake but the vague sense of a more powerful creature, the serpent, surrounded by a mist of fear and wonderment.

These qualities are dominant in a dream I have experienced often through my life, for reasons that will soon become clear.

I find myself in a locality that is wooded and aquatic, silent and drawn wholly in shades of gray. As I walk into this somber environment I am gripped by an alien feeling. The terrain before me is mysterious, on the rim of the unknown, at once calm and forbidding. I am required to be there but in the dream state cannot grasp the reasons. Suddenly, the Serpent appears. It is not an ordinary, literal reptile, but much more, a threatening presence with unusual powers. It is protean in size and shape, armored, irresistible. The poisonous head radiates a cold, inhuman intelligence. While I watch, its muscular coils slide into the water, beneath prop roots, and back onto the bank. The Serpent is somehow both the spirit of that shadowed place and the guardian of the passage into deeper reaches. I sense that if I could capture or control or even just evade it, an indefinable but great change would follow. The anticipation of that transformation stirs old and nameless emotions. The risk is also vaguely felt, like that emanating from a knife blade or high cliff. The Serpent is life-promising and life-threatening, seductive and treacherous. It now slips close to me, turning importunate, ready to strike. The dream ends uneasily, without clear resolution.

The snake and the serpent, flesh-and-blood reptile and demonic dream-image, reveal the complexity of our relation to nature and the fascination and beauty inherent in all organisms. Even the deadliest and most repugnant creatures are endowed with magic in the human mind. Human beings have an innate fear of snakes; more precisely, they have an innate propensity to learn such fear quickly and easily past the age of five. The images they build out of this peculiar mental set are both powerful and ambivalent, ranging from terror-stricken flight to the experience of power and male sexuality. As a consequence the serpent has become an important part of cultures around the world.

There is a highly complex principle to consider here, one that extends well beyond the ordinary concerns of psychoanalytic reasoning about sexual symbols. Life of any kind is infinitely more interesting than almost any conceivable variety of inanimate matter. The latter is valued chiefly to the extent that it can be metabolized into live tissue, accidentally resembles it, or can be fashioned into a useful and properly animated artifact. No one in his right mind looks at a pile of dead leaves in preference to the tree from which they fell.

What is it exactly that binds us so closely to living things? The biologist will tell you that life is the self-replication of giant molecules from lesser chemical fragments, resulting in the assembly of complex organic structures, the transfer of large amounts of molecular information, ingestion, growth, movement of an outwardly purposeful nature, and the proliferation of closely similar organisms. The poet-in-biologist will add that life is an exceedingly improbable state, metastable, open to other systems, thus ephemeral—and worth any price to keep.

Certain organisms have still more to offer because of their special impact on mental development. In 1984, in a book titled *Biophilia*, I suggested that the urge to affiliate with other forms of life is to some degree innate. The evidence for the proposition is not strong in a formal scientific sense: the subject has not been studied enough in the scientific manner of hypothesis, deduction, and experimentation to let us be certain about it one way or the other. Nevertheless the biophilic tendency is so clearly evinced in daily life and so widely distributed as to deserve serious attention. It unfolds in the predictable fantasies and responses of individuals from early childhood onward. It cascades into repetitive patterns of culture across most or all societies, a consistency often noted in the literature of anthropology. These processes appear to be part of the programs of the brain. They are marked by the quickness and decisiveness with which we learn particular things about certain kinds of plants and animals. They are too consistent to be dismissed as the result of purely historical events etched upon a mental blank slate.

Perhaps the most bizarre of the biophilic traits is awe and veneration of the serpent. The dreams from which the dominant images arise are known to exist in all societies whose mental life has been studied. At least 5 percent of the people at any given time remember experiencing them, while many more would probably do so if they recorded their waking impressions over several months. The images described by urban New Yorkers are as detailed and emotional as those of Australian aboriginals and Zulus. In all cultures the serpents are prone to be mystically transfigured. The Hopi know Palulukon, the water serpent, a benevolent but frightening godlike being. The Kwakiutl fear the *sisiutl*, a three-headed serpent with both human and reptile faces, whose appearance in dreams presages insanity or death. The Sharanahua of Peru summon reptile spirits by taking hallucinogenic drugs and stroking their faces with the severed tongues of snakes. They are rewarded with dreams of brightly colored boas, venemous snakes, and lakes teeming with caimans and anacondas. Around the world serpents and snakelike creatures are the dominant elements of dreams in which animals of any kind appear. They are recruited as the animate symbols of power and sex, totems, protagonists of myths, and gods.

These cultural manifestations may seem at first detached and mysterious, but there is a simple reality behind the ophidian archetype that lies within the experience of ordinary people. The mind is primed to react emotionally to the sight of snakes, not just to fear them but to be aroused and absorbed in their details, to weave stories about them. This distinctive predisposition played an important role in an unusual experience of my own, a childhood encounter with a large and memorable snake, a creature that actually existed.

I grew up in the panhandle of northern Florida and the adjacent counties of Alabama. Like most boys in that part of the country set loose to roam the woods, I enjoyed hunting and fishing and made no clear distinction between these activities and life at large. But I also cherished natural history for its own sake and decided very early to become a biologist. I had a secret ambition to find a Real Serpent, a snake so fabulously large or otherwise different that it would exceed the bounds of imagination, let alone existing fact.

Certain circumstances encouraged this adolescent fantasy. First of all, I was an only child with indulgent parents, encouraged to develop my own interests and hobbies, however farfetched; in other words, I was spoiled. Second, the physical surroundings inclined youngsters toward an awe of nature. Four generations earlier, that part of the country had been covered by a wilderness as formidable in some respects as the Amazon. Dense thickets of cabbage palmetto descended into meandering spring-fed streams and cypress sloughs. Carolina parakeets and ivory-billed woodpeckers flashed overhead in the sunlight, and wild turkeys and passenger pigeons still counted as game. On soft spring nights after heavy rains a dozen varieties of frogs croaked, rasped, bonged, and trilled their love songs in mixed choruses. Much of the Gulf Coast fauna derived from species that had spread north from the tropics over millions of years and adapted to the warm local temperate conditions. Columns of miniature army ants, close replicas of the large marauders of South America, marched mostly unseen at night over the forest floor. *Nephila* spiders the size of saucers spun webs as wide as garage doors across the woodland clearings.

From the stagnant pools and knothole sinks, clouds of mosquitoes rose to afflict the early immigrants. They carried the Confederate plagues, malaria and yellow fever, which periodically flared into epidemics and reduced the populations along the coastal lowlands. This natural check is one of the reasons the strip between Tampa and

Pensacola remained sparsely settled well into the twentieth century and why even today, long after the diseases have been eradicated, it is still the relatively natural "other Florida."

Snakes abounded. The Gulf Coast has a greater variety and denser populations than almost any other place in the world, and they are frequently seen. Striped ribbon snakes hang in gorgonlike clusters on branches at the edge of ponds and streams. Poisonous coral snakes root through the leaf litter, their bodies decorated with warning bands of red, yellow, and black. They are easily confused with their mimics, the scarlet kingsnakes, banded in a different sequence of red, black, and yellow. The simple rule recited by woodsmen is: "Red next to yellow will kill a fellow, red next to black is a friend of Jack." Hognoses, harmless thick-bodied sluggards with upturned snouts, are characterized by an unsettling resemblance to venomous African gaboon vipers and a habit of swallowing toads live. Pygmy rattlesnakes two feet long contrast with diamondbacks of seven feet or more. Watersnakes are a herpetologist's medley distinguished by size, color, and the arrangement of body scales, encompassing ten species of *Natrix*, *Seminatrix*, *Agkistrodon*, *Liodytes*, and *Farancia*.

Of course limits to the abundance and diversity exist. Because snakes feed on frogs, mice, fish, and other animals of similar size, they are necessarily scarcer than their prey. You can't just go out on a stroll and point to one individual after another. An hour's careful search will often turn up none at all. But I can testify from personal experience that on any given day you are ten times more likely to meet a snake in Florida than in Brazil or New Guinea.

There is something oddly appropriate about the abundance of snakes. Although the Gulf wilderness has been largely converted into macadam and farmland, and the sounds of television and company jets are heard in the land, a remnant of the old rural culture remains, as if the population were still pitted against the savage and the unknown. "Push the forest back and fill the land" remains a common sentiment, the colonizer's ethic and tested biblical wisdom (the very same that turned the cedar groves of Lebanon into the barrens they are today). The prominence of snakes lends symbolic support to this venerable belief.

In the back country during a century and half of settlement, the common experience of snakes was embroidered into the lore of serpents. Cut off a rattlesnake's head, one still hears, and it will live on until sundown. If a snake bites you, open the puncture wounds with a knife and wash them with kerosene to neutralize the poison (if there are any survivors of this cure, I have never met them). If you believe with all your heart in Jesus, you can hang rattlers and copperheads around your neck without fear. If one strikes you just the same, accept it as a sign from the Lord and find peace in whatever follows. The hognose snake, on the other hand, is always death in the shape of a slithery S. Those who get too close to one will have venom sprayed in their eyes and be blinded; the very breath from the snake's skin is lethal. This species is the beneficiary of its dreadful legend: I have never heard of any being killed.

Deep in the woods live creatures of startling power. (*That* is what I most wanted to hear.) Among them is the hoop snake. Skeptics, who used to be found hunkered down in a row along the county courthouse balustrade on a Saturday morning, say it is only mythical; on the other hand it might be the familiar coachwhip racer turned vicious by special circumstances. Thus transformed, it puts its tail in its mouth and rolls down hills at great speed to attack its terrified victims. Then there are reports of the occasional true monsters: a giant snake believed to live in a certain swamp (used to be there anyway, even if no one's seen it in recent years); a twelve-foot diamondback rattler a farmer killed on the edge of town a few years back; some unclassifiable prodigy recently glimpsed as it sunned itself along the river's edge.

It is a wonderful thing to grow up in southern towns where animal fables are taken half-seriously, breathing into the adolescent mind a sense of the unknown and the possibility that something extraordinary might be found within a day's walk of where you live. No such magic exists in the environs of Schenectady, Liverpool, and Darmstadt, and for all children dwelling in such places where the options have finally been closed, I feel a twinge of sadness. I found my way out of Mobile, Pensacola, and Brewton to explore the surrounding woods and swamps with a languorous intensity. I formed the habit of quietude and concentration into which I still pass my mind during field excursions, having learned to summon the old emotions as part of the naturalist's technique.

Some of these feelings must have been shared by my friends. In the mid-1940s during the hot season between spring football practice and the regular schedule of games in the fall, working on highway cleanup gangs and poking around outdoors were about all we had to do. But there was some difference: I was hunting snakes with passionate commitment. On the Brewton High School football team of 1944–45 most of the players had nicknames leaning toward the infantilisms and initials favored by southerners: Bubba Joe, Flip, A. J., Sonny, Shoe, Jimbo, Junior, Snooker, Skeeter. As the underweight third-string left end, allowed to play only in the fourth quarter when the foe had been crushed beyond any hope of recovery, mine was Snake. But although I was inordinately proud of this measure of masculine acceptance, my main hopes and energies had been invested elsewhere. There are an incredible forty species of snakes native to that region, and I managed to capture almost all of them.

One kind became a special target just because it was so elusive: the glossy watersnake *Natrix rigida.* The adults lay on the bottom of shallow ponds well away from the shore and pointed their heads out of the alga-green water in order to breathe and to scan the surface in all directions. I waded out toward them very carefully, avoiding the side-to-side movements to which snakes are most alert. I needed to get within three or four feet in order to manage a diving tackle, but before I could cover the distance they always pulled their heads under and slipped silently away into the opaque depths. I finally solved the problem with the aid of the town's leading slingshot artist, a taciturn loner my age, proud and quick to anger, the sort of boy who in an earlier time might have distinguished himself at Antietam or Shiloh. Aiming pebbles at the heads of the snakes, he was able to stun several long enough for me to grab them underwater. After recovering, the captives were kept for a while in homemade cages in our backyard, where they thrived on live minnows placed in dishes of water.

Once, deep in a swamp miles from home, half lost and not caring, I glimpsed an unfamiliar brightly colored snake disappearing down a crayfish burrow. I sprinted to the spot, thrust my hand after it, and felt around blindly. Too late: the snake had squirmed out of reach into the lower chambers. Only later did I think about the possibilities: suppose I had succeeded and the snake had been poisonous? My reckless enthusiasm did catch up with me on another occasion when I miscalculated the reach of a pygmy rattlesnake, which struck out faster than I thought possible and hit me with startling authority on the left index finger. Because of the small size of the reptile, the only results were a temporarily swollen arm and a fingertip that still grows a bit numb at the onset of cold weather.

I found my Serpent on a still July morning in the swamp fed by the artesian wells of Brewton, while working toward higher ground along the course of a weed-choked stream. Without warning a very large snake crashed away from under my feet and plunged into the water. Its movement especially startled me because so far that day I had encountered only modestly proportioned frogs and turtles silently tensed on mudbanks and logs. This snake was more nearly my size as well as violent and noisy—a colleague, so to speak. It sped with wide body undulations to the center of the shallow watercourse and came to rest on a sandy riffle. Though not quite the monster I had envisioned, it was nevertheless unusual, a water moccasin *(Agkistrodon piscivorus)*, one of the poisonous pit vipers, more than five feet long with a body as thick as my arm and a head the size of a fist. It was the largest snake I had ever seen in the wild. I later calculated it to be just under the published size record for the species. The snake now lay quietly in the shallow clear water completely open to view, its body stretched along the fringing weeds, its head pointed back at an oblique angle to watch my approach. Moccasins are like that. They don't always keep going until they are out of sight, in the manner of ordinary watersnakes. Although no emotion can be read in the frozen half-smile and staring yellow cat's eyes, their reactions and posture make them seem insolent, as if they see their power reflected in the caution of human beings and other sizable enemies.

I moved through the snake handler's routine: pressed the snake stick across the body in back of the head, rolled it forward to pin the head securely, brought one hand around to grasp the neck just behind the swelling masseteric muscles, dropped the stick to seize the body midway back with the other hand, and lifted the entire animal clear of the water. The technique almost always works. The moccasin, however, reacted in a way that took me by surprise and put my life in immediate danger. Throwing its heavy body into convulsions, it twisted its head and neck slightly forward through my gripped fingers, stretched its mouth wide open to unfold the inch-long fangs and expose the dead-white inner lining in the intimidating "cottonmouth" display. A fetid musk from its anal glands filled the air. At that moment the morning heat became more noticeable, the episode turned manifestly frivolous, and at last I wondered what I was doing in that place alone. Who would find me? The snake began to turn its head far enough to clamp its jaws on my hand. I was not very strong for my age, and I was losing control. Without thinking I heaved the giant out into the brush, and this time it thrashed frantically away until it was out of sight and we were rid of each other.

I sat down and let the adrenaline race my heart and

bring tremors to my hand. How could I have been so stupid? What is there in snakes anyway that makes them so repellent and fascinating? The answer in retrospect is deceptively simple: their ability to remain hidden, the power in their sinuous limbless bodies, and the threat from venom injected hypodermically through sharp hollow teeth. It pays in elementary survival to be interested in snakes and to respond emotionally to their generalized image, to go beyond ordinary caution and fear. The rule built into the brain in the form of a learning bias is: become alert quickly to any object with the serpentine gestalt. *Overlearn* this particular response in order to keep safe.

Other primates have evolved similar rules. When guenons and vervets, the common monkeys of the African forest, see a python, cobra, or puff adder, they emit a distinctive chuttering call that rouses other members in the group. (Different calls are used to designate eagles and leopards.) Some of the adults then follow the intruding snake at a safe distance until it leaves the area. The monkeys in effect broadcast a dangerous-snake alert, which serves to protect the entire group and not solely the individual who encountered the danger. The most remarkable fact is that the alarm is evoked most strongly by the kinds of snakes that can harm them. Somehow, apparently through the routes of instinct, the guenons and vervets have become competent herpetologists.

The idea that snake aversion is inborn in man's relatives

is supported by studies of rhesus macaques, the large brown monkeys of India and surrounding Asian countries. When adults see a snake of any kind, they react with the generalized fear response of their species. They variously back off and stare (or turn away), crouch, shield their faces, bark, screech, and twist their faces into the fear grimace—lips retracted, teeth bared, and ears flattened against the head. Monkeys raised in the laboratory without previous exposure to snakes show the same response to them as those brought in from the wild, though in weaker form. During control experiments designed to test the specificity of the response, the rhesus failed to react to other, nonsinuous objects placed in their cages. It is the form of the snake and perhaps also its distinctive movements that contain the key stimuli to which the monkeys are innately tuned.

Grant for the moment that snake aversion does have a hereditary basis in at least some kinds of nonhuman primates. The possibility that immediately follows is that the trait evolved by natural selection. In other words, individuals who respond leave more offspring than those who do not, and as a result the propensity to learn fear quickly spreads through the population—or, if it was already present, is maintained there at a high level.

How can biologists test such a proposition about the origin of behavior? They turn natural history upside down: they search for species historically free of forces in the environment believed to favor the evolutionary change, to see if in fact the organisms do *not* possess the trait. Lemurs, primitive relatives of monkeys, offer such an inverted opportunity. They are indigenous inhabitants of Madagascar, where no large or poisonous snakes exist to threaten them. Sure enough, lemurs presented with snakes in captivity fail to display anything resembling the automatic fear responses of the African and Asian monkeys. Is this adequate proof? In the chaste idiom of scientific discourse, we are permitted to conclude only that the evidence is "consistent with the proposition." Neither this nor any comparable hypothesis can be settled by a single case. Only further examples can raise confidence in it to a level beyond potential challenge by determined skeptics.

Another line of evidence comes from studies of the chimpanzee, a species thought to have shared a common ancestor with prehumans as recently as 5 million years ago. Chimps raised in the laboratory become apprehensive in the presence of snakes, even if they have had no previous experience. They back off to a safe distance and follow the intruder with a fixed stare while alerting companions with the *Wah!* warning call. More important, the response becomes gradually more marked during adolescence.

This last quality is especially interesting because human beings pass through approximately the same developmental sequence. Children under five years of age feel no special anxiety over snakes, but later they grow increasingly wary. Just one or two mildly bad experiences, such as the sight of a garter snake writhing away in the grass, having a rubber model thrust at them by a playmate, or hearing a counselor tell scary stories at the campfire, can make children deeply and permanently fearful. The pattern is unusual if not unique in the ontogeny of human behavior. Other common fears, notably of the dark, strangers, and loud noises, start to wane after seven years of age. In contrast, the tendency to avoid snakes grows stronger with time. It is possible to turn the mind in the opposite direction, to learn to handle snakes without apprehension or even to like them in some special way, as I did—but the adaptation takes a special effort and is usually a little forced and self-conscious. The special sensitivity is just as likely to lead to full-blown ophidiophobia, the pathological extreme in which the mere appearance of a snake brings on a feeling of panic, cold sweat, and waves of nausea. I have witnessed these events.

At a campsite in Alabama, on a Sunday afternoon, a four-foot-long black racer glided out from the woods across the clearing and headed for the high grass along a nearby stream. Children shouted and pointed. A middle-aged woman screamed and collapsed to the ground sobbing. Her husband dashed to his pickup truck to get a shotgun. But black racers are among the fastest snakes in the world, and this one made it safely to cover. The onlookers probably did not know that the species is nonvenomous and harmless to any creature larger than a cotton rat.

Halfway around the world, in the village of Ebabaang in New Guinea, I heard shouting and saw people running down a path. When I caught up with them they had formed a circle around a small brown snake that was essing leisurely across the front yard of a house. I pinned the snake and carried it off to be preserved in alcohol for the museum collections at Harvard. This seeming act of daring earned either the admiration or the suspicion of my hosts—I couldn't be sure which. The next day children followed me around as I gathered insects in the nearby forest. One brought me an immense orb-weaving spider gripped in his fingers, its hairy legs waving and the evil-looking black fangs working up and down. I felt panicky and sick. It so happens that I suffer from mild arachnophobia. To each his own.

Why should serpents have such a strong influence during mental development? The direct and simple answer is that throughout the history of mankind a few kinds have been a major cause of sickness and death. Every continent except Antarctica has poisonous snakes. Over large stretches of Asia and Africa the known death rate from snakebite is 5 persons per 100,000 each year or higher. The local record is held by a province in Burma, with 36.8 deaths per 100,000 a year. Australia has an exceptional abundance of deadly snakes, a majority of which are relatives of the cobra. Among them the tiger snake is especially feared for its large size and tendency to strike without warning. In South and Central America live the bushmaster, fer-de-lance, and jaracara, among the largest and most aggressive of the pit vipers. With backs colored like rotting leaves and fangs long enough to pass through a human hand, they lie in ambush on the floor of the tropical forest for the small warm-blooded animals that constitute their major prey. Few people realize that a complex of dangerous snakes, the "true" vipers, are still relatively abundant throughout Europe. The common adder *Viperus berus* ranges to the Arctic Circle. The number of people bitten in such improbable places as Switzerland and Finland is still high enough, running into the hundreds annually, to keep outdoorsmen on a sort of yellow alert. Even Ireland, one of the few countries in the world lacking snakes altogether (thanks to the last Pleistocene glaciation and not Saint Patrick), has imported the key ophidian symbols and traditions from other European cultures and preserved the fear of serpents in art and literature.

Here, then, is the sequence by which the agents of nature appear to have been translated into the symbols of culture. For hundreds of thousands of years, time enough for the appropriate genetic changes to occur in the brain, poisonous snakes have been a significant source of injury and death to human beings. The response to the threat is not simply to avoid it, in the way that certain berries are recognized as poisonous through a process of trial and error. People also display the mixture of apprehension and morbid fascination characterizing the nonhuman primates. They inherit a strong tendency to acquire the aversion during early childhood and to add to it progressively, like our closest phylogenetic relatives, the chimpanzees.

The mind then adds a great deal more that is distinctively human. It feeds upon the emotions to enrich culture. The tendency of the serpent to appear suddenly in dreams, its sinuous form, and its power and mystery are the natural ingredients of myth and religion.

Consider how sensation and emotional states are elaborated into stories during dreams. The dreamer hears a distant thunderclap and changes an ongoing episode to end with the slamming of a door. He feels a general anxiety and is transported to a schoolhouse corridor, where he searches for a classroom he does not know in order to take an examination for which he is unprepared. As the sleeping brain enters its regular dream periods, marked by rapid eye movement beneath closed eyelids, giant fibers in the lower brainstem fire upward into the cortex. The awakened mind responds by retrieving memories and fabricating stories around the sources of physical and emotional discomfort. It hastens to recreate the elements of past real experience, often in a jumbled and antic form. And from time to time the serpent appears as the embodiment of one or more of these feelings. The direct and literal fear of snakes is foremost among them, but the dream-image can also be summoned by sexual desire, a craving for dominance and power, and the apprehension of violent death.

We need not turn to Freudian theory in order to explain our special relationship to snakes. The serpent did not originate as the vehicle of dreams and symbols. The relation appears to be precisely the other way around and correspondingly easier to study and understand. Humanity's concrete experience with poisonous snakes gave rise to the Freudian phenomena after it was assimilated by genetic evolution into the brain's structure. The mind has to create symbols and fantasies from something. It leans toward the most powerful preexistent images or at least follows the learning rules that create the images, including that of the serpent. For most of this century, perhaps overly enchanted by psychoanalysis, we have confused the dream with the reality and its psychic effect with the ultimate cause rooted in nature.

Among prescientific people, whose dreams are conduits to the spirit world and snakes a part of ordinary experience, the serpent has played a central role in the building of culture. There are magic incantations for simple protection, as in the hymns of the Atharva Veda: "With my eye do I slay thy eye, with poison do I slay thy poison. O Serpent, die, do not live; back upon thee shall thy poison turn."

"Indra slew thy first ancestors, O Serpent," the chant continues, "and since they are crushed, what strength forsooth can be theirs?" And so the power can be controlled and even diverted to human use through iatromancy and the casting of magic spells. Two serpents entwine the caduceus, which was first the winged staff of Mercury as messenger of the gods, then the safe-conduct pass of ambassadors and heralds, and finally the universal emblem of the medical profession (by whom it was confused with the staff of Asclepius, Greek god of medicine, which was entwined with a single serpent).

Balaji Mundkur has shown how the inborn awe of snakes matured into rich productions of art and religion around the world. Serpentine forms wind across stone carvings from paleolithic Europe and are scratched into mammoth teeth found in Siberia. They are the emblems of power and ceremony for the shamans of the Kwakiutl, the Siberian Yakut and Yenisei Ostyak, and many tribes of Australian aboriginals. Stylized snakes have often served as the talismans of the gods and spirits who bestow fertility: Ashtoreth of the Canaanites, the demons Fu-Hsi and Nu-kua of the Han Chinese, and the powerful goddesses Mudammā and Manasā of Hindu India. The ancient Egyptians venerated at least thirteen ophidian deities ministering to various combinations of health, fecundity, and vegetation. Prominent among them was the triple-headed giant Nehebkau, who traveled widely to inspect every part of the river kingdom. Amulets in gold inscribed with the sign of a cobra god were placed in Tutankhamen's funeral wrappings. Even the scorpion goddess Selket bore the title "mother of serpents." Like her offspring she prevailed simultaneously as a source of evil, power, and goodness.

The Aztec pantheon was a phantasmagoria of monstrous forms among whom serpents were given pride of place. The calendrical symbols included the ophidian *olin nahui* and *cipactli*, the earth crocodile that possessed a forked tongue and rattlesnake's tail. The rain god Tlaloc consisted in part of two coiled rattlesnakes whose heads met to form the god's upper lip. *Coatl*, serpent, is the dominant element in the names of Aztec divinities. Coatlicue was a threatening chimera of snake and human parts, Cihuacoatl the goddess of childbirth and mother of the human race, and Xiuhcoatl the fire serpent over whose body fire was rekindled every fifty-two years to mark a major division in the religious calendar. Quetzalcoatl, the plumed serpent with a human head, reigned as god of the morning and evening stars and thus of death and resurrection. As inventor of the calendar, deity of books and learning, and patron of the priesthood, he was revered in the schools where nobles and priests were taught. His reported departure over the eastern horizon upon a raft of snakes must have been the occasion of consternation for the intellectuals of the day, something like the folding of the Guggenheim Foundation.

Contradictory ophidian images were a feature of ancient Greek religion as well. Among the early forms of Zeus was the serpent Meilichios, at once god of love, gentle and responsive to supplication, and god of vengeance, whose sacrifice was offered at night. Another great serpent protected the lustral waters at the spring of Ares. He coexisted with the Erinyes, avenging spirits of the underworld so horrible that they could not be pictured in early mythology. Euripides depicted them as serpents in his *Iphigeneia in Tauris:* "Dost see her, her the Hades-snake who gapes / To slay me, with dread vipers, open-mouthed?"

Slyness, deception, malevolence, betrayal, the implicit threat of a forked tongue flicking in and out of the mask-like head, all qualities tinged with miraculous powers to heal and guide, forecast and empower, became the serpent's prevailing image in western cultures. The serpent in the Garden of Eden, appearing as in a dream to serve as Judaism's evil Prometheus, gave humankind knowledge of good and evil and with it the burden of original sin, in return for which God ordained:

I will put enmity between you and the woman,
between your brood and hers.
They shall strike at your head,
and you shall strike at their heel.

To summarize the relation between human and snake: life becomes part of us. Culture transforms the snake into the serpent, a far more potent creation than the literal reptile. Culture, as a product of the mind, can be interpreted as an image-making machine that recreates the outside world through symbols arranged into maps and stories. But the mind does not have the capacity to grasp reality in its full chaotic richness; nor does the body last long enough for the brain to process information piece by piece like an all-purpose computer. Rather, consciousness races ahead to master certain kinds of information with enough efficiency to survive. It submits to a few biases easily, while automatically avoiding others. A great deal of evidence has accumulated in genetics and physiology to show that the controlling devices are biological in nature, built into the sensory apparatus and brain by particularities in cellular architecture.

The combined biases are what we call human nature. The central tendencies, exemplified so strikingly in fear and veneration of the serpent, are the wellsprings of culture. Hence simple perceptions yield an unending abundance of images with special meaning while remaining true to the forces of natural selection that created them.

How could it be otherwise? The brain evolved into its present form over a period of about 2 million years, from the time of *Homo habilis* to the late Stone Age of *Homo sapiens*, during which people existed in hunter-gatherer bands in intimate contact with the natural environment. Snakes mattered. The smell of water, the hum of a bee, the directional bend of a plant stalk mattered. The naturalist's trance was adaptive: the glimpse of one small animal hidden in the grass could make the difference between eating and going hungry in the evening. And a sweet sense of horror, the shivery fascination with monsters and creeping forms that so delights us today even in the sterile hearts of the cities, could keep you alive until the next morning. Organisms are the natural stuff of metaphor and ritual. Although the evidence is far from all in, the brain appears to have kept its old capacities, its channeled quickness. We stay alert and alive in the vanished forests of the world.

IN PRAISE OF SHARKS

TWO BIOCHEMISTS ARE AT a scientific conference in the Caribbean, sitting on a dock and dangling their feet in the water as they discuss the day's proceedings. Suddenly a dark shadow passes below, followed by a swirl of water, and the left leg of one of the men is jerked downward.

"My God!" shouts the stricken scientist. "A shark just bit off my toe."

"Good Lord, no!" exclaims the other, peering into the water. "Which one?"

"How should I know?" responds the first, after a moment's reflection. "When you've seen one shark, you've seen them all."

I often use this apocryphal little story in class to illustrate the difference between scientists who, like the biochemists, stress general principles of form and function—properties of *the* shark, for example—and those who emphasize the diversity of life. The latter, the evolutionary biologists, are more interested in how species arise and how diversity is maintained through time.

In fact there are about 350 species of shark by recent

count—excluding their close relatives, the rays and skates—and they differ from one another in ways almost as extreme as it's possible to imagine. To get a feel for this diversity, we can start with a creature appropriately called the garbage can of the sea, the tiger shark *(Galeocerdo cuvieri)*. Tiger sharks, which can attain a length of 20 feet and weigh almost a ton, often patrol refuse-filled harbors, where they're attracted to virtually anything containing animal protein or, for that matter, almost anything at all. Stomachs of captured specimens have contained fish, boots, beer bottles, bags of potatoes, coal, dogs, even parts of human beings. One giant yielded this haul: three overcoats, a raincoat, a driver's license, one cow's hoof, the antlers of a deer, twelve undigested lobsters, and a chicken coop with feathers and bones inside. It's not surprising that swimmers occasionally get nabbed, but it can truthfully be said that nothing personal was intended. The tiger shark is simply a big eater, not especially an enemy of man.

Then there's the cookie-cutter shark *(Isistius brasiliensis)*, an 18-inch-long parasite of porpoises, whales, and such large fish as bluefin tuna. (A parasite is a predator that eats its prey in units of less than one, without killing it, at least not right away.) This little fish has a curving row of very large teeth on its lower jaw, which it thrusts into the bodies of its victims and twists to slice out one- to two-inch-wide conical plugs of skin and flesh. For many years the circular scars on porpoises and whales were a mystery, thought by some to be caused by bacterial infection or invertebrate parasites, until the habits of cookie-cutters were discovered in 1971. These little sharks have been known to attack nuclear submarines, taking bites out of the rubber coat of the sonar domes.

Tiny as the cookie-cutters are, they're not the smallest sharks. That distinction may belong to an obscure species, the dwarf shark *(Squaliolus laticaudus)*, which reaches a known maximum length of only one foot. At the opposite extreme is the whale shark *(Rhincodon typus)*, the largest fish in the world; whale sharks 60 feet long and weighing more than 10 tons have been reported. But all that bulk is no menace to humans, or to any other creatures larger than the small fishes and planktonic animals on which the whale shark feeds. *Rhincodon* has evolved to resemble the baleen whale in its mode of feeding, to say nothing of size. It swims with deliberation, often just below the surface, running large volumes of water through its mouth to seine out its small prey. Brave swimmers have dived down alongside whale sharks, taken hold of their dorsal fins, and hitched a ride.

Sharks, as a group, display one of the best cases in the living world of the phenomenon evolutionary biologists call adaptive radiation: the proliferation of species individually specialized to fill very different ecological niches. Birds provide a familiar example; they have diversified mightily to produce predators, scavengers, insect feeders, seed-eating finches, ostriches and other large flightless species, amphibious penguins (land and water), triphibious auks (land, water, and air), nectar-feeding hummingbirds and sunbirds, and other types specialized in anatomy and behavior to a similar degree. The diversification clearly reduces competition among the member species and permits more of them to be packed into local habitats—that is, allows more to live together for long periods without becoming extinct. In an archipelago such as the Galápagos or Hawaii, so remote from the mainland that it can be reached by only a few species over long periods, successful colonists can diversify to fill many of the traditional major niches more quickly than elsewhere. Even more spectacular that Darwin's famous Galápagos finches are the Hawaiian honeycreepers, comprising more than twenty species, which evolved from a single goldfinchlike species that arrived millions of years ago from either Asia or North America.

On a global scale, sharks have achieved one of the greatest adaptive radiations. The 350 species fill most of the major niches occupied by all the kinds of fishes put together, plus those of the whales and porpoises. In addition to the familiar tiger sharks and others of conventional appearance and behavior, there are gulper sharks, bramble sharks, wobbegongs, mandarin dogfish, spurdogs, saw sharks, probeagles, goblin sharks, crocodile sharks, sleeper sharks, pygmy sharks, and many others.

Think of almost anything a medium to large bony fish might do, and you'll find one or more species of shark that does it about as well, or perhaps better. On the deep ocean floor live the eel-like frilled sharks, with the same oversized mouth and needle-sharp teeth that characterize deep-water predatory fish. Thousands of feet above them cruise blue sharks, black-tipped sharks, and other streamlined species, which, like mackerel and bluefish, are beautifully built for pursuit and maneuverability. On the continental shelves are found the angel sharks, sluggish forms with square flat bodies superficially resembling those of rays and torpedo fish; and saw sharks, whose grotesque snouts, lined by outward-projecting teeth, make them hard to distinguish from the "true" sawfish.

Major diversification of this kind often produces basic types found in no other group. The thresher shark is among the more impressive examples. It herds fish and squid into bunches with swift charges, then stuns its victims with strikes of its long, whiplike tail. Because of this peculiarity, fishermen often hook threshers in the tail instead of the mouth. At the opposite extreme are the wobbegongs of the western Pacific. These sharks are club-shaped, with fleshy barbels arrayed along their mouths and the sides of their heads like mustaches and sideburns. Their mottled coloration enables them to blend into the ocean floor and has inspired their other name, carpet shark. Torpid in disposition, they "walk" along the bottom with their pectoral fins. Wobbegongs are dangerous to human beings. When stepped on they sometimes flip around and seize their provokers with needle-sharp teeth, hanging on with a bulldog grip. Such attacks are no trivial matter; the spotted wobbegong reaches a length of 10 feet.

I have a criterion for the attainment of a *real* adaptive radiation: it has been reached when at least one species specializes in feeding on other members of the same group. Army ants, for example, eat other kinds of ants, king snakes other snakes. The sharks have arrived as well: near the mouth of the Mississippi, bull sharks, growing to 500 pounds, feed preferentially on a variety of smaller sharks. In deeper waters, tiger sharks and hammerheads take the same prey on a more casual basis.

The ultimate product of all this evolution, in my opinion, is the great white shark, *Carcharodon carcharias.* It has rightly been called a top carnivore, a killing machine, the last free predator of man. The great white is by far the largest flesh-eating fish on Earth. It's known with certainty to reach 21 feet and 7,300 pounds; there are some unproved claims of 26 feet and 9,000 pounds. The belly of the shark is white, its back slate gray to black. Its teeth, each forming a saw-edged equilateral triangle, stand in rows along the edge of its mouth, and regenerate easily when broken off. The front of the head, the snout, is tapered into a cone, a conspicuous feature giving rise to the alternative Australian name of white pointer. (With characteristic pungency, Australians also call this species white death.) The mouth of the shark is usually set in a little clown's grin, agape and with teeth on display, bringing in a flow of water back across the gills in the fashion of a ramjet. The great white is warm-bodied; it maintains a body temperature well above that of the surrounding water. Perhaps as a result, it's distributed throughout the colder waters of most of the oceans of the world and forages from the surface down to 4,300 feet.

Carcharodon carcharias consumes a wide range of bony fish, other sharks, sea turtles, and marine mammals such as porpoises, seals, and sea lions. So much does the mature great white favor sea mammals that these otherwise solitary creatures sometimes gather close to seal and sea lion rookeries, such as those at California's Farallon Islands and at Dangerous Reef off South Australia. The great white is dangerous to people simply because it doesn't always make a clear distinction between a seal and a human swimmer.

A great white shark is alerted to prey at great distances by smell, and then moves closer to investigate. In moderately clear water it can see someone swimming or paddling on a surfboard from 20 to 40 feet away. Whereas other sharks are likely to move in with caution, swimming around the prey and poking it before attacking, the great white goes straight to the kill. It rushes upward toward the prey, and at the last moment rolls its eyes backward, extends its upper jaw (by lifting its snout and head), drops its lower jaw, and takes a bite, all with flashing speed—usually in less than a second, according to Tim Tricas and John McCosker of San Francisco's Steinhard Aquarium, who have filmed white sharks feeding off southern Australia. Then the fish often swims a short distance away and lets its victim bleed to death. This last habit has allowed many swimmers to survive, at least when others are nearby to come to their rescue. It also helps the rescuers, who are seldom attacked even when towing the victim ashore.

For a time some shark experts thought this special behavior of great whites indicated that they were man-biters rather than man-eaters, and that perhaps human flesh or the neoprene of diving suits was distasteful to the fish. Other scientists speculated that the shark bit only to defend its territory. McCosker dismisses both theories and cites as evidence the manner in which the great white attacks humans—from beneath and behind, just as it attacks its usual prey, seals and sea lions. Why? Because, as McCosker explains, the great white has learned over millions of years that as soon as a seal or sea lion spots it, it's out of luck, or at least a meal: the agile sea mammals can easily circle out of the clumsy shark's path and elude its powerful jaws. So the great white must rely on stealth.

Unfortunately it hasn't yet learned, if it ever will, to make subtle distinctions. In the past few decades, divers in their rubberized suits have come to look more and more like seals and sea lions. The great white looks up, sees what seems like the silhouette of its familiar prey, takes a bite, waits for its victim to bleed to death, then finishes the job.

What does it feel like to be the prey of a creature 20 times your size? Frank Logan, while fishing for abalone at the southern end of California's Bodega Bay in 1968, felt a strange numbing pressure on his left side. He turned to find a large part of his trunk in the mouth of a shark whose body "went clear out of sight in the murky water." He had been hit by an 18- to 20-foot great white shark. Here's how Logan recalled the experience: "It pushed me sideways through the water, maybe 10 to 20 feet, I'm not sure. But I could feel the water eddying past my body; I just went limp and played dead. I knew if the shark shook me, it would tear me apart. Everything happened so fast, I didn't have time to be scared. I said to myself, 'Let me go, please let me go!'—I don't know how long it was, maybe 20 seconds. And then it let me go." Logan escaped with the help of friends, but it took 200 stitches to close all the wounds that stretched in a 20-inch arc across his body.

Although the great white has a way of life that makes it particularly menacing to human beings, and although no scuba diver in his right mind would stay in the water with one, except perhaps behind the bars of a strong steel cage, it doesn't often attack humans. Over the past 375 years, New England has lost only one person to a great white, sixteen-year-old Joseph Troy Jr., killed while swimming in Buzzards Bay, Massachusetts, on July 25, 1936. Even along the California coast, where attack rates are among the highest in the world, there is on average only one fatality every eight years. In contrast, 10 to 20 white sharks are killed by fishermen every year. Since the sharks are much less numerous to begin with than people who use the water, they clearly are getting the worst of the deal, although the balance could change. McCosker believes that with coastal mammals like seals and otters under federal protection and increasing in population, the number of great whites will also increase, and along with them the number of attacks on humans, especially off California and Oregon.

Sharks have persisted in one form or another since the Devonian period, some 400 million years ago, and hence are more than 100 times older than anything that could remotely be called human. They've remained fairly abundant throughout all this time, except for one short dip during the early part of the age of dinosaurs, and have been increasing in diversity and possibly also in abundance for the past 50 million years. Theirs is a record rivaled only by cockroaches, scorpions, and a very few other groups of animals.

Why have sharks been so successful? Zoologists aren't sure, but they point to several traits that seem to contribute to superior adaptability. Fertilization is internal, and in most species the young are born alive and are able to swim away on their own power immediately. Sharks are able to feed heavily when they succeed in capturing prey and then fast for weeks at a time, living on food stored in their livers. In fact huge livers are as important a part of the biology of sharks as their gill slits and deciduous teeth.

Consisting mostly of oil, the livers account for 10 to 25 percent of the weight of the fish.

If sheer bulk and power are the criteria, the greatest fish story ever told involved a whale shark. In 1959 G. S. Illugason of the United Nations Food and Agriculture Organization and two assistants were teaching new techniques to thirteen Indian fishermen in the Arabian Sea just west of Mangalore. They were working in two steel-hulled boats, 27 and 32 feet long, secured together by a rope. Spotting a giant whale shark passing, Illugason decided to try for it with the only equipment available, a 30-inch unbarbed hook and a 2-inch manila line. He caught the hook in the dorsal fin of the fish, which continued on its course, pulling the two boats at a steady 5 knots. After three hours the shark tired enough for the men to secure it with two more lines and a steel wire wound around its dorsal fin. Seven hours after the first encounter, the shark was pulled ashore. Its length was 32 feet and its estimated weight 5 tons, more than some of the local fishermen would catch in a lifetime.

The conservation of shark species hasn't begun, although a case can already be made for the protection of such conspicuous and harmless giants as the whale and basking sharks. Our problem is ignorance. Very little is known about the great majority of the 350 species, usually no more than approximately where they live, a little of their anatomy, and something of their food habits. But I confess I like that state of affairs. It stirs the blood to realize that large wild animals still roam free in an unexplored part of the world. And for scientists and naturalists the unknown has always been much more interesting than the sampled, the photographed, and the measured. Songs unheard are sweeter far . . .

In 1976, 500 feet from the surface in 15,000 feet of water northeast of Oahu, something became entangled in a parachute deployed as a sea anchor by a U.S. Navy research vessel. When hauled up onto the slanted stern with the aid of rollers used to retrieve torpedoes, the creature proved to be a 14-foot, 1,650-pound shark of a completely new kind. It had an extraordinarily big head and a huge mouth with which it had been filter-feeding on euphausiid shrimp when it was caught by the anchor. Astonished scientists christened it the megamouth shark or, more formally, *Megachasma pelagios.* In November 1984 another specimen was caught near Santa Catalina Island, off southern California, and several others have turned up in the western Pacific. What else is swimming down there?

Sharks are part of the world in which we evolved, and therefore part of us. They're instilled in our culture as a mirror of our most deeply rooted anxieties and fears. Heedless of our concern about them, they live on as they have lived for hundreds of millions of years, symbols of mystery and a still untamed world.

IN THE COMPANY OF ANTS

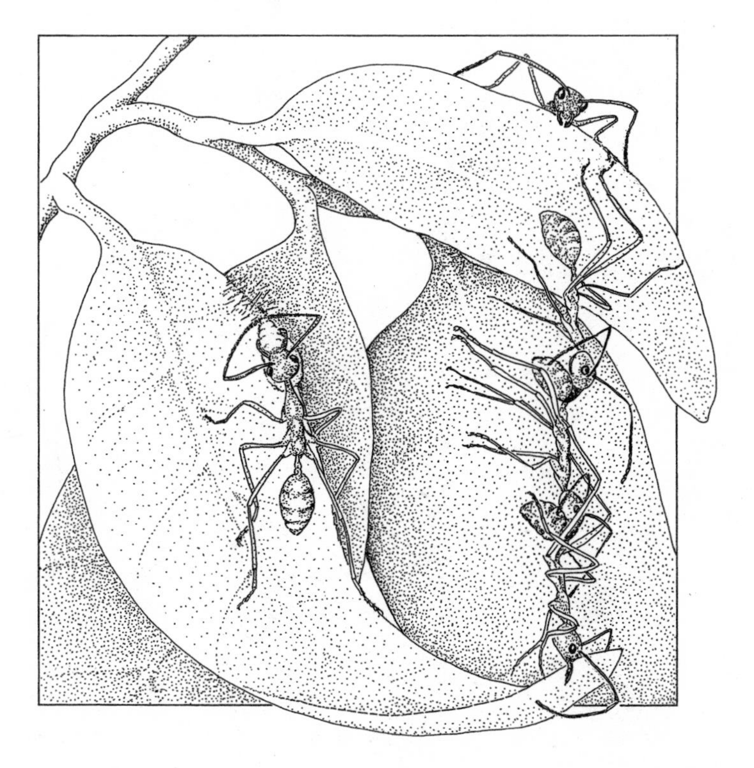

THE QUESTION I'M ASKED most often about ants is "What do I do about the ones in my kitchen?" And my answer is always the same: "Watch where you step." Be careful of little lives. Feed them crumbs of coffeecake. They also like bits of tuna and whipped cream. Get a magnifying glass. Watch them closely. And you will be as close as any person may ever come to seeing social life as it might evolve on another planet. The evolutionary line that gave rise ultimately to ants and other social insects separated more than 600 million years ago from the line that gave rise to human beings. Insect social systems are completely independent of our own and differ from it in many profound ways. They're another grand experiment in evolution for our delectation. The study of their unique traits has already proved very rewarding in several fields of biology.

At present there are about 9,500 described species of ants; this is the number so far given a scientific name. I'd venture a guess that there are in actual existence two or three times that many, and there is immense diversity within this group of hymenopterous insects. A colony of the world's smallest ant could dwell comfortably inside the braincase of the world's largest ant. One genus of ants that I've been studying, *Pheidole*, contains 285 named species from the New World alone. In the collection at Harvard's Museum of Comparative Zoology I have about 600 species; in other words, some 315 are new to science. More pour in from collectors every few months.

Ants are the dominant little-sized organisms of the planet—that is, intermediate in size between bacteria and elephants. My rough estimate is that at any given moment there are about 10^{15}, or a million billion, ants in the world. In terms of overall biomass, measured as dry weight, they are truly formidable. For example, in forests near Manaus, in the central Brazilian Amazon, ants and termites together make up more than one-quarter of the biomass—which includes everything from very small worms and other invertebrates to the largest mammals. Ants alone weigh four times as much as the birds, amphibians, reptiles, and mammals combined. This proportion of ants is approached or exceeded in most other major types of land habitat around the world. When we consider insect biomass alone, we find that the ants and termites, the most highly social of all organisms, plus the social wasps and social bees, which rival them in colonial organization, make up about 80 percent of the biomass. These insects dominate the insect world from the Arctic Circle to Tierra del Fuego and Tasmania. In fact, ants are the principal predators of small animals roughly their own size. They are the "cemetery squad," scavenging and removing the corpses of more than 90 percent of the small animals. They are movers and enrichers of the earth, more so than the earthworms. Indeed, although the social insects as a group make up only 2 percent of all of the known described species of insects in the world, they probably make up most of the biomass.

Ants have been present for about 100 million years, since the middle of the Cretaceous period of the Mesozoic era, and have been among the most abundant insects for the past 50 million years. In 1967 two colleagues and I at Harvard had the privilege of describing the first ants of the Mesozoic era, and they proved to be true missing links. The specimens, which were found in New Jersey by amateur fossil collectors and which we named *Sphecomyrma* ("wasp ant"), combine in a remarkable manner traits of the presumed ancestral wasp and modern ants. Subsequently the Russians came up with a host of new fossils of roughly the same age.

How have ants managed to stay on top of things for a period fifty times longer than the entire history of human beings and their immediate ancestors? I'll give you what I believe to be the correct short answer, and then I'll expand on the theme implicit in it.

Ants and other social insects are dominant because their social organization gives them competitive superiority over solitary insects. Wherever you go in the world, from rain forest to desert, social insects occupy the center—the stable, resource-rich parts of the environment. Solitary insects, although they also exist in great abundance, are specialists of the fringes—the ephemeral part of the habitat. They are concentrated on outer foliage, deep in wood, in tiny crevices of the soil, and in other sites not preempted by the social insects. A colony of ants can be regarded as a kind of superorganism—a gigantic, amoebalike entity that blankets the foraging field, collecting food and launching forays to engage enemies before they can approach the nest. At the same time they care for the queen and the immature ants—ranging from eggs through larvae to pupae—that are sequestered with her in the nest. They accomplish all these things with high efficiency by a division of labor. Most important, they do them simultaneously. No task is left undone for more than a very short time. No enemy is left unchallenged; no hapless caterpillar fallen from a tree is left uncollected. Also, individuals are able to risk or even to sacrifice their lives in suicidal ventures on behalf of the colony without greatly reducing its productivity. Through close identity with their common mother, the queen, they are able to take far greater risks, in a Darwinian sense, than are solitary insects—and they often do so through mass defense and recruitment to the battlefield, using tactics whose sophistication is worthy of a von Clausewitz. Ant societies are the most warlike of all known animal groups, solitary or social. Most species of ants engage in frequent territorial battles, during which kamikazelike assaults by sterile workers turn the tide. In the deserts of the Southwest, for example, scouts of the genus *Dorymyrmex*, upon discovering a nest of their rivals in the genus *Myrmecocystus*, recruit fellow colony members, who then surround the nest entrance, take bits of gravel to the edge of the nest, and drop them in. Any *Myrmecocystus* that continue to resist eventually get buried under the rubble, which at least temporarily closes off their exit to the outside. And in the Malaysian rain forest, worker ants of certain species of *Camponotus* have grotesquely hypertrophied paired glands that open out at the base of the mandibles and fill a large part of the body. These receptacles are loaded with a sticky toxic chemical. When confronted by enemies, and under extreme duress, the ants are able to contract their abdominal muscles and explode in the face of the enemy, rather like walking grenades. One of these ants can trade its life for those of several enemies. In Darwinian terms this is an excellent tactic.

Another reason that ant social life succeeds is that colonies work to maintain the nest as a climate-controlled factory within a fortress. Inside it the queen and the nurse-workers busily raise young and rapidly increase the population. The nest itself is constructed to hold off enemies. It is protected by the often very aggressive work force, which in many species includes a specialized soldier caste. These ants also control large areas around the nest, from which they harvest food. Furthermore, they are able to bequeath the nest, which is very expensive to produce in terms of energy, together with the territory, to later generations. In southern Finland the nests of mound-building ants are as much as 2 meters high and are thought to be hundreds of years old. Such nests, together with the ants' social system, confer the capacity to build large, dense populations that dominate the environment.

All of these marvelously complex activities are instinctive and gene-driven. They could not possibly be learned or otherwise "culturally transmitted."

Let me make these social principles more explicit by describing two examples of what might be termed the high civilizations of the ant world. Both are species of which I have firsthand experience. A great deal of my research on the first species was conducted with Bert Hölldobler, now at the University of Würzburg.

The African and Asian weaver ants (genus *Oecophylla*), which date back at least 50 million years, to the late Eocene epoch, live in the tree crowns of tropical forests. These ants dominate much of the canopy, not only because of their large size as individuals, but also because of their large populations. Mature colonies contain more than 200,000 workers. A remarkable system of communication allows the colony to occupy the tops of several trees—an area spanning tens of thousands of square feet. And like the Roman empire at its peak, this territory is connected tightly by networks of trails. The ants also maintain garrisons from which workers go out to collect prey and defend the nest. They make part of their domiciles, which include tunnels and pavilions, from silk. They also use silk to bind leaves and twigs together.

There is only one mother queen in each colony. She is attended by her daughters. The colony is an all-female society. Males are reared and kept in the nests for short periods and are present only to inseminate a virgin queen on nuptial flights. They die immediately after performing this duty. The workers of this species are divided into two castes according to size. The major workers take care of most of the quotidian tasks of the colony, including nurturing the queen, hunting, and building and defending the nest. The minor workers specialize for the most part in caring for the young; they are the nursing caste.

Each colony has hundreds of pavilions, clusters of leaves held together with webs of silk. One pavilion may contain many thousands of workers. The pavilions near the periphery of the colony's territory are occupied primarily by the oldest workers, the age group that generally comprises the warriors in ant societies. These are the ants most prone to risk their lives in defense of the colony. Thus one basic difference between human and ant societies is that whereas human beings send their young men to war, ants send their old women.

Another difference is that whereas we orient ourselves and communicate chiefly through sight and sound, ants do so largely through taste and smell. In most species the body of each worker contains between ten and twenty exocrine glands that release chemical secretions to the outside of the body in one manner or another, to be smelled or tasted by fellow members of the colony. These secretions variously alarm, recruit, identify members of the colony, identify caste, and so on.

The weaver ants have what may be the most complex chemical system known in the animal kingdom. The workers have no fewer than five separate recruitment systems, which are distinguished by the context in which the pheromones are delivered, as well as by which tactile signals the ants present with them (for example, how they tap, approach, rush at, or stand above other ants). The combined signals inform the ants of the situation and cause them to respond accordingly. Translated into English, the five recruitment systems of the weaver ant are "enemy close by," "enemy at long distance," "new territory discovered that we can reach," "new suitable site on which to build a pavilion," and "food."

Perhaps even more remarkable is the way in which weaver ants build a pavilion, and on which their common name is based. Their labor is highly specialized and coordinated. First, mass formations of workers apply the pressure required to fold leaves and bring them together so that they can be bound by silk. The ants form a living chain by grasping each other about the waist; the one at the end grabs a leaf to pull it and curl it over. If a single chain does not suffice, the ants form themselves into sheets of living chains to make the vegetation conform to their needs. When the leaves are properly lined up, specialized workers bring out their immature sisters, small, grublike larvae that are at a late stage in their development, and employ them as living shuttles. A worker puts a larva's head down on the spot where it wants a silk thread to be released and touches the larva with its antennae, giving it a precise signal to release a strand. As the larva lets out a strand, the ant pulls it sideways to the edge of another leaf. This process is repeated literally thousands of times, until the larva has no silk left with which to spin a cocoon. But it doesn't matter; the naked pupae are protected in the fearsome colonies and come to maturity anyway.

The second exemplar of high civilization is the leafcutter ant (genus *Atta*) of the New World tropics. There are actually about a dozen species, but all apparently have similar habits in maintaining agricultural states of imperial proportions. The colonies subsist almost entirely on a fungus raised on leaves and other vegetation, which are harvested fresh. They also feed somewhat on plant sap. Only one kind of fungus grows with them, and it is totally dependent on the ants.

The colony is created by a single queen, a giant insect about half the size of your thumb. While a virgin, she still has wings. She leaves the mother nest on nuptial flights. In the air, she and others—her sisters and other queens from other colonies, swarming by the millions—meet the males that also fly out for the single act that justifies their brief existence. While still in the air she mates with five or more males, storing all the sperm obtained in a small elastic bag next to the oviduct. They are sufficient in number to fertilize all the eggs required to make about 150 million daughter workers, of which between 2 and 3 million are alive at any given time over the 10- to 15-year life of the colony. Next the queen settles to the ground, and her dry and membranous wings drop off painlessly along a special abscission line. She digs a hole in the ground and prepares to lay eggs and start a colony. But wait, you may say—how is she going to start the fungus garden? Before leaving the mother nest, the queen has carefully gathered up strands of fungus there and tucked them into a special pocket at the base of her mouth. Now she regurgitates the strands, lays eggs, and uses the eggs and the feces she produces to start a fungus garden on the floor of the nest.

An intricate division of labor based on caste, which in turn is based on wide variations in size (head widths ranging from 0.8 to more than 5 millimeters), allows the colonies of this species to create and maintain their agricultural economies. The ants form an incessant assembly line that moves the processing of leaves and fungus from the largest workers to progressively smaller ones. Operating in the thousands, the largest workers cut leaves, flowers, and stems in a stereotypical manner as much as 100 meters from home. Carrying the excised pieces like umbrellas over their heads, they run back to the nest along odor trails consisting of a dimethylpyrazine laid down from the poison gland at the tip of the abdomen. So powerful is this substance that only a few molecules are sufficient to stimulate an ant. The chemists who identified the chemical estimate that if it could be paid out with theoretical maximum efficiency, one gram could lead a column of ants twice around the world.

The energy expenditure of these largest workers is equally impressive. I used to be a track statistics buff, and for idle amusement I converted into human terms the speed at which they move during their leaf-transporting trips. If one of these ants were a six-foot-tall person, it would be running along those trails of pyrazine at a pace of about 3:45 minutes per mile. That's about the current human world record. At the end of the trail, after running roughly the distance of a marathon, it would pick up a load of 300 pounds or more and carry it home at the slightly slower pace of 4 minutes per mile. Upon reaching the nest, it would climb down through the galleries and chambers of the nest for a distance of up to one mile before depositing its leaf load.

Now back to the assembly line. When the fragments come into the nest, they are turned over to a class of slightly smaller workers, which cut the leaves into pieces about a millimeter across. These pieces are taken over by still-smaller workers, who chew them into little wads and defecate on them, thus delivering digestive enzymes onto the bits of leaves. These enzymes, which are found in the fungus on which the ants subsist, somehow travel through the gut of the ant without being digested. Yet-smaller workers now use the little globs of chewed and treated leaf material to build a spongelike structure on top of the fungus garden. Even-smaller workers then take tufts of the growing fungus from elsewhere and implant it in the globs. The smallest workers of all (which constitute the most numerous caste) tend the fungus, weeding out alien fungus species and performing intricate gardening maneuvers. The fungus possesses tasty little swollen tips that can be plucked like vegetables off the growing mass and eaten.

My laboratory studies revealed that as the colony grows older, and as its population expands from just a few workers to nearly 100,000, the size-frequency distribution changes in a quite predictable "programmed demography." The death and birth schedules of the various castes almost always progress in the same way. Remarkably, the frequency distribution of new workers produced by the queen when she first builds the underground chamber exactly spans the minimal size variation needed to create the colony's assembly line. If the queen were to make the single mistake of raising an overly large worker, absorbing an excessive amount of food, too few workers of other sizes could be raised to produce a complete assembly line, and the colony would die. This societal-level demographic effect appears quite clearly to have been molded by natural selection.

In many ways ants challenge our ingenuity and command our attention. Their social order is different from our own in almost every key respect. They seized control of a large part of the terrestrial environment long before the first primates, let alone the first human beings, walked the earth. For most of 100 million years they've imposed a deep imprint on the remainder of terrestrial life. In terms of their great success and longevity they have a great deal to teach us—not by example, surely, but by the illumination of the interlocking principles that join sociobiology to ecology and the study of evolution.

ANTS AND COOPERATION

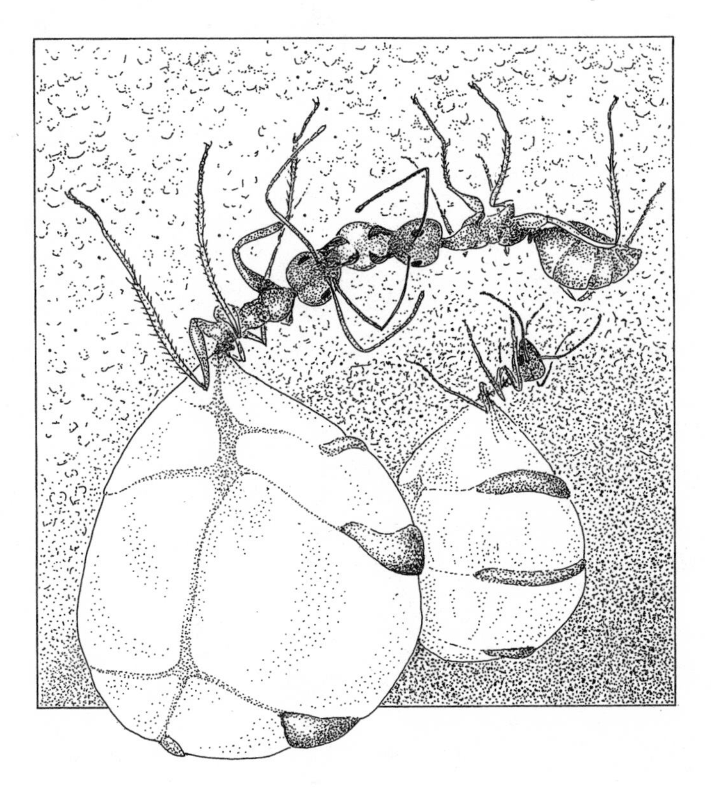

IN A SUNLIT CLEARING ON Mexico's Yucatán peninsula, a giant black worker ant, a female like all members of her caste, leaves her earthen nest and climbs a nearby shrub to a glistening cluster of dew. She's on a mission of survival for herself and her clan. Opening her mandibles, she collects a droplet of dew, then returns to the nest. After pausing at the entrance to allow another worker to drink some of the water, she descends through vertical galleries until she reaches the brood chambers where the colony's immature young are kept. There she daubs part of her burden onto a cocoon and passes the rest to a thirsty larva.

During dry periods the ant colony, like those of all social insects, is in mortal danger of desiccation. Many of the workers make repeated trips back and forth to sources of water wherever it can be found. Some share the water with nestmates, while others place the drops directly on the ground inside the brood chambers, keeping the soil and air moist and thus protecting their young sisters during their most vulnerable period of development.

Through such cooperative behavior, the colony is able to survive and grow even during the hardest of times.

These Gunga Dins of the insect world are of a species of giant tropical stinging ant, *Pachycondyla villosa.* Half an inch long fully grown, they can inflict a throbbing pain to humans that lasts for days. But it's their water sharing that especially interests scientists. In my view, the sharing of food and water is a more important component of advanced social behavior than dominance, leadership, or any other kind of interaction. When sharing is extended beyond offspring to include siblings and less closely related individuals—in other words, when it becomes truly altruistic—it tightens social bonds and leads to the evolution of some of the most complex forms of communication in the animal kingdom.

The development of similar patterns may have played a key role in the evolution of human social behavior. Limited fossil evidence indicates that 2 million or more years ago, the earliest "true" men, *Homo habilis*, lived at African campsites to which they carried food and distributed it to others. Anthropologists think this arrangement, which persisted and intensified throughout prehistory, favored complex communications, long-term reciprocal agreements, and thus ultimately a uniquely rich social existence. Today virtually all cultures employ food sharing as part of their bonding rituals and rites of passage.

Procedures of sharing are also at the heart of social life among insects. Water transport by the *Pachycondyla* ants, for example, is only one aspect of their system of communal feeding. The workers also gather drops of nectar and carry them home between their mandibles. The liquid is distributed to nestmates, which store it temporarily by holding it between their own mandibles. What began as a single large drop when the forager first entered the nest finally ends up being carried around by ten or more workers. The ants also hunt other insects, which they transport home to be torn into small pieces and shared among colony members.

Bert Hölldobler found that drop-carrying is widespread within the group of primitive ants to which *Pachycondyla* belongs. These mostly tropical ants, constituting the subfamily Ponerinae, originated as far back as the late Mesozoic era, 70 million years ago. Almost all are stinging insects that capture live prey. And in a pattern paralleling that of mammals, the species whose workers hunt in groups are also by and large the ones that possess the most complex societies and modes of communication. A few go so far as to organize formidable raiding parties that overwhelm colonies of termites and other species of ants.

Compared to such sophisticated insect behavior, drop-carrying is rudimentary. Other workers beg part of the drops by lightly drumming on the heads of the carriers with their antennae and legs. This combination of signals is identical with that used by a wide variety of primitive ants to recruit nestmates to new nest locations and food sources. During the evolution of ants the recruitment function evidently appeared first and was only later expanded to trigger the sharing of liquid.

The vast majority of the 9,500 known species of ants living in the world today have evolved an even cleverer way of distributing liquid: storing it inside the bodies of the workers themselves. Water, nectar, and sometimes dissolved fats are passed down the esophagus to the crop, a muscular organ that expands and contracts like a small balloon. When a worker imbibes deeply, its crop swells and expands its entire abdomen. You can observe this phenomenon yourself by offering drops of sugar water or honey to ants wandering around the house and letting them drink their fill. When they return home (usually right away and in a straight line), they will pass some of the water to other members of the colony by mouth-to-mouth regurgitation.

Here the story gets still more interesting and significant. Some years ago Hölldobler found that workers of evolutionarily more advanced ant species begged food from nestmates by drumming their antennae and front legs on the food-bearer's labium, the hinged plate at the lower surface of the mouth that functions roughly like a lower lip. Stimulated in this way, the workers automatically regurgitated a drop of liquid up from their crop out to the space between their mandibles. Hölldobler was able to induce the response himself by touching a hair to the labium. He also found that certain parasitic beetles living with ants got free meals by closely imitating the begging movements of their hosts. The ants didn't seem to notice that the beetles were radically different in shape and never gave any of the food back.

At about the same time that Hölldobler conducted his studies, Thomas Eisner of Cornell and I used radioactively labeled sugar water to trace the distribution of liquefied food through a colony of common black ants *(Formica subsericea)* by the regurgitation process. We found that portions of the food brought in by a single worker reached every other worker in the colony within 24 hours, after prolonged bouts of reciprocal feeding. Within a week all the colony members were carrying approximately the same quantity of the radioactive material. We had confirmed the opinion held by earlier entomologists that the crop serves as a "social stomach." That is, what a worker holds in her crop at any given moment is approximately what the rest of the colony possesses. So when the colony as a whole is hungry, the same is true of each of the foraging workers to a closely similar degree. When the colony requires a particular nutrient, the foragers look for it—they have no need to be told.

In the principal deserts of the world, a few species of *Camponotus* and other kinds of ants have carried the exchange of liquid to its logical extreme by evolving a special food-storage caste. Certain large workers, while still young, are given extra rations of sugary liquid, with the result that their abdomens swell into large translucent bubbles. Once transformed, they remain mostly in one spot for the rest of their lives. Only when an enemy breaks into their living quarters or when nest conditions become uncomfortable do they drag themselves slowly over the ground in order to move to a new site. These repletes, as they're called by entomologists, are living storage pots of liquid nourishment. During the rainy season, when the air around the nests is relatively cool and food abundant, the foraging workers fill the repletes to capacity by regurgitation. During the hottest, driest months, it's the storage ants' turn to regurgitate as the foragers and others draw on the supply.

The elaboration of liquid and food sharing in ants, from drop-carrying to regurgitation, is important in binding the colony members together and coordinating their activities. Yet despite all our efforts over many years of research, those of us who study social insects have never been able to find a command center. No individual—not even the queen, an oversized creature concerned mainly with reproduction—lays plans for the colony as a whole. No one, for example, designates which ants will become members of the storage caste and which will specialize as guardians of the nest. Instead, the activity of an ant colony or beehive is the summation of a vast number of personal decisions by individual ants. When everyone has roughly the same stomach content, individual decisions become similar, and a more harmonious form of mass action is possible.

Each worker ant has a brain consisting of about a million nerve cells. The average human being, who is one million times as heavy as an ant, has a brain of approximately 100 billion nerve cells. Insects are therefore constitutionally not very bright, and they must rely on automatic guidance systems such as uniform food sharing to keep their colonies running. This is the reason why most kinds of ants, though very impressive in their social attainments, have apparently changed little since the age of the dinosaurs. It's also why they may outlast our own fractious, impatient species.

THE PATTERNS

OF NATURE

ALTRUISM AND AGGRESSION

During wars of this century, a large percentage of Congressional Medals of Honor have been awarded to men who threw themselves on top of grenades to shield comrades, aided the rescue of others from battle at the price of certain death to themselves, or made other, often carefully considered but extraordinary, decisions that led to the same end. Such altruistic self-sacrifice is the ultimate act of courage and emphatically deserves the country's highest honor. It is also only the extreme act that lies beyond the innumerable smaller performances of kindness and giving that bind societies together. One is tempted to leave the matter there, to accept altruism as simply the better side of human nature. Perhaps, to put the best possible construction on the matter, conscious altruism is a transcendental quality that distinguishes human beings from animals. But scientists are not accustomed to declaring any phenomenon off limits, and in the past two decades there has been a renewed interest in analyzing such forms of social behavior in greater depth and as objectively as possible.

Much of the new effort falls within a discipline called sociobiology, which is defined as the systematic study of the biological basis of social behavior in every kind of organism, including man, and is being pieced together with contributions from biology, psychology, and anthropology. There is nothing new about analyzing social behavior, and even the word "sociobiology" has been around for years. What is new is the way facts and ideas are being extracted from their traditional matrix of psychology and ethology (the natural history of animal behavior) and reassembled in compliance with the principles of genetics and ecology.

In sociobiology, there is a heavy emphasis on the comparison of societies of different kinds of animals and of man, not so much to draw analogies (these have often been dangerously misleading, as when aggression is compared directly in wolves and in human beings) but to devise and to test theories about the underlying hereditary basis of social behavior. With genetic evolution always in mind, sociobiologists search for the ways in which the myriad forms of social organization adapt particular species to the special opportunities and dangers encountered in their environment.

A case in point is altruism. I doubt if any higher animal, such as a hawk or a baboon, has ever deserved a Congressional Medal of Honor according to the ennobling criteria used in our society. Yet minor altruism does occur frequently, in forms instantly understandable in human terms, and is bestowed not just on offspring but on other members of the species as well. Certain small birds—robins, thrushes, and titmice, for example—warn others of the approach of a hawk. They crouch low and emit a distinctive thin, reedy whistle. Although the warning call has acoustic properties that make it difficult to locate in space, to whistle at all seems at the very least unselfish; the caller would be wiser not to betray its presence but rather to remain silent and let someone else fall victim.

When a dolphin is harpooned or otherwise seriously injured, the typical response of the remainder of the school is to desert the area immediately. But sometimes they crowd around the stricken animal and lift it to the surface, where it is able to continue breathing air. Packs of African wild dogs, the most social of all carnivorous mammals, are organized in part by a remarkable division of labor. During the denning season some of the adults, usually led by a dominant male, are forced to leave the pups behind in order to hunt for antelopes and other prey. At least one adult, normally the mother of the litter, stays behind as a guard. When the hunters return, they regurgitate pieces of meat to all that stayed home. Even sick and crippled adults are benefited, and as a result they are able to survive longer than would be the case in less generous societies.

Other than man, chimpanzees may be the most altruistic of all mammals. Ordinarily chimps are vegetarians, and during their relaxed foraging excursions they feed singly in the uncoordinated manner of other monkeys and apes. But occasionally the males hunt monkeys and young baboons for food. During these episodes the entire mood of the troop shifts toward what can only be characterized as a manlike state. The males stalk and chase their victims in concert; they also gang up to repulse any of the victims' adult relatives that oppose them. When the hunters have dismembered the prey and are feasting, other chimps approach to beg for morsels. They touch the meat and the faces of the males, whimpering and *hoo*ing gently, and hold out their hands—palms up—in supplication. Sometimes the meat eaters pull away in refusal or walk off. But often they permit the other animal to chew directly on the meat or to pull off small pieces with its hands. On several occasions chimpanzees have even been observed to tear off pieces and drop them into the outstretched hands of others—an act of generosity unknown in other monkeys and apes.

Adoption is also practiced by chimpanzees; Jane Goodall has observed three cases at the Gombe Stream National Park in Tanzania. All involved orphaned infants taken over by adult brothers and sisters. It is of considerable interest, for more theoretical reasons to be discussed shortly, that the altruistic behavior was displayed by the closest possible relatives rather than by experienced females with children of their own, females who might have supplied the orphans with milk and more adequate social protection.

Despite a fair abundance of such examples among vertebrates, it is only in the lower animals, and in the social insects particularly, that we encounter altruistic suicide comparable to the human level. A large proportion of the members of colonies of ants, bees, and wasps are ready to defend their nests with insane charges against intruders. This phenomenon explains why people move with circumspection around honeybee hives and yellowjacket burrows but can afford to relax near the nests of solitary species such as sweat bees and mud daubers.

The social stingless bees of the tropics swarm over the heads of humans who venture too close, locking their jaws so tightly onto tufts of hair that their bodies pull loose from their heads when they are combed out. Some species pour a burning glandular secretion onto the skin during these sacrificial attacks; in Brazil they are called *cagafogos* ("fire defecators"). The great entomologist William Morton Wheeler described an encounter with the "terrible bees," during which they removed patches of skin from his face, as the worst experience of his life.

Honeybee workers have stings lined with reversed barbs like those on fishhooks. When a bee attacks an intruder at the hive, the sting catches in the skin; as the bee moves away, the sting remains embedded, pulling out the entire venom gland and much of the viscera with it. The bee soon dies, but its attack has been more effective than if it had withdrawn the sting intact: the venom gland continues to leak poison into the wound, while a bananalike odor emanating from the base of the sting incites other members of the hive to launch kamikaze attacks of their own at the same spot. From the point of view of the colony as a whole, the suicide of an individual accomplishes more than it loses. The total worker force consists of 20,000 to 80,000 members, all sisters born from eggs laid by the mother queen. Each bee has a natural life span of only about 50 days, at the end of which it dies of old age. So to give a life is only a little thing, with no loss of genes in the process.

My favorite example among the social insects is provided by an African termite with the orotund technical name *Globitermes sulfureus.* Members of this species' soldier caste are quite literally walking bombs. Huge paired glands extend from their heads back through most of their bodies. When they attack ants and other enemies, they eject through their mouths a yellow glandular secretion that congeals in the air and often fatally entangles both the soldiers and their antagonists. The spray appears to be powered by contractions of the muscles in the abdominal wall. Sometimes the contractions become so violent that the abdomen and gland explode, spraying the defensive fluid in all directions.

A shared capacity for extreme sacrifice does not mean that the human mind and the "mind" of an insect (if such exists) work alike. But it does mean that the impulse need not be ruled divine or otherwise transcendental, and we are justified in seeking a more conventional biological explanation. Such an explanation immediately poses a basic problem: fallen heroes don't have any more children. According to the narrow mode of Darwinian natural selection, self-sacrifice results in fewer descendants, and the genes, or basic units of heredity, that allow heroes to be created can be expected to disappear gradually from the population. Because people who are governed by selfish genes appear to prevail over those with altruistic genes, there should be a tendency over many generations for selfish genes to increase in number and for the human population as a whole to become less and less capable of responding in an altruistic manner.

How can altruism persist? In the case of the social insects, there is no doubt at all. Natural selection has been broadened to include a process called kin selection. The self-sacrificing termite soldier protects the rest of the colony, including the queen and king, which are the soldier's parents. As a result the soldier's more fertile brothers and sisters flourish, and *they* multiply the altruistic genes that are shared with the soldier by close kinship. One's own genes are multiplied by the greater production of nephews and nieces. Has the capacity for altruism also evolved in human beings through kin selection? Do the emotions we feel, which on occasion in exceptional individuals culminate in total self-sacrifice, stem ultimately from hereditary units that were implanted by the favoring of relatives during a period of hundreds or thousands of generations? This explanation gains some strength from the circumstance that during most of human history the social unit was the immediate family and a tight network of other close relatives. Such exceptional cohesion, combined with a detailed awareness of kinship made possible by high intelligence, might explain why kin selection has been more forceful in human beings than in monkeys and other mammals.

To anticipate a common objection raised by many social scientists and others, let me grant at once that the intensity and form of altruistic acts are to a large extent culturally determined. Human social evolution is obviously more cultural than genetic. Nevertheless, the underlying emotion, powerfully manifested in virtually all human societies, is considered by sociobiologists to evolve through genes. Although this hypothesis does not therefore account for differences among societies, it could explain why humans differ from other mammals and why, in one narrow aspect, they more closely resemble social insects.

In cases in which sociobiological explanations can be tested and proved true, they will, at the very least, provide perspective and a new sense of philosophical ease about human nature. I believe that they may also have an ultimately moderating influence on social tensions. Consider the case of homosexuality. Homophiles are typically rejected in our society because of a narrow and unfair biological premise made about them: their sexual preference does not produce children; therefore, they cannot be natural. Nevertheless, homosexuals *can* replicate genes by kin selection, provided they are sufficiently altruistic toward kin.

It is not inconceivable that in the early, hunter-gatherer period of human evolution, and perhaps even later, homosexuals regularly served as a partly sterile caste, enhancing the lives and reproductive success of their relatives by a more dedicated form of support than would have been possible if they produced children of their own. If such combinations of interrelated heterosexuals and homosexuals regularly left more descendants than similar groups of pure heterosexuals, the capacity for homosexual development would remain prominent in the population as a whole.

Supporting evidence for this new kin-selection hypothesis does not exist; the idea has not even been examined critically. But the fact that it is internally consistent and can be squared with the results of kin selection in other kinds of organisms should give us pause before labeling homosexuality an illness. If the hypothesis is correct, we can expect homosexuality to decline over many generations, since the extreme dispersal of family groups in modern industrial societies leaves fewer opportunities for preferred treatment of relatives. The labor of homosexuals is spread more evenly over the population at large, and the narrower form of Darwinian natural selection turns against the duplication of genes favoring this kind of altruism.

A moderating role of modern sociobiology is also possible in the interpretation of aggression, the behavior at the opposite pole from altruism. To cite aggression as a form of social behavior seems paradoxical; considered in terms of individual acts, it is more accurately identified as antisocial behavior. But when viewed in a social context, it seems to be one of the most important and widespread organizing techniques. Animals use it to stake out their own territories and to establish their rank in the pecking orders. And because members of one group often cooperate for the purpose of directing aggression at competitor groups, altruism and hostility have come to be opposite sides of the same coin.

Konrad Lorenz, in his celebrated 1966 book *On Aggression*, argued that human beings share a general instinct for aggressive behavior with animals, and that this instinct must somehow be relieved, if only through competitive sport. Erich Fromm, in *The Anatomy of Human Destructiveness* (1973), took the still dimmer view that man's behavior is subject to a unique death instinct that often leads to pathological aggression beyond that encountered in animals. Both of these interpretations are essentially wrong. A close look at aggressive behavior in a variety of animal societies, many of which have been carefully studied only since the time Lorenz drew his conclusions, shows that aggression occurs in a myriad of forms and is subject to rapid evolution.

We commonly find one species of bird or mammal to be highly territorial, employing elaborate, aggressive displays and attacks, while a second, otherwise similar, species shows little or no territorial behavior. In short, the case for a pervasive aggressive instinct does not exist.

The reason for the lack of a general drive seems quite clear. Most kinds of aggressive behavior are perceived by biologists as particular responses to crowding in the environment. Animals use aggression to gain control over necessities—usually food or shelter—that are in short supply or likely to become so at some time during the life cycle. Many species seldom or never run short of these necessities; rather, their numbers are controlled by predators, parasites, or emigration. Such animals are characteristically pacific in their behavior toward one another.

Mankind happens to be one of the aggressive species. But we are far from being the most aggressive. Recent studies of hyenas, lions, and langur monkeys have disclosed that under natural conditions these animals engage in lethal fighting, infanticide, and even cannibalism at a rate far above that found in humans. When a count is made of the number of murders committed per thousand individuals per year, human beings are well down the list of aggressive creatures, and I am fairly confident that this would still be the case even if our episodic wars were included. Hyena packs engage in deadly pitched battles that are virtually indistinguishable from primitive human warfare. Hans Kruuk of Oxford University has described the action of two packs in the Ngorongoro Crater:

> *The two groups mixed with an uproar of calls, but within seconds the sides parted again and the Mungi hyenas ran away, briefly, pursued by the Scratching Rock*

> *hyenas, who then returned to the carcass. About a dozen of the Scratching Rock hyenas, though, grabbed one of the Mungi males and bit him wherever they could, especially in the belly, the feet and the ears. The victim was completely covered by his attackers, who proceeded to maul him for about 10 minutes while their clan fellows were eating the wildebeest. The Mungi male was literally pulled apart, and when I later studied the injuries more closely, it appeared that his ears were bitten off and so were his feet and testicles, he was paralyzed by a spinal injury, had large gashes in the hind legs and belly, and subcutaneous hemorrhages all over . . . The next morning, I found a hyena eating from the carcass and saw evidence that more had been there; about one-third of the internal organs and muscles had been eaten. Cannibals!*

Alongside ants, which conduct assassinations, skirmishes, and pitched battles as routine business, humans are all but tranquil pacifists. Ant wars are especially easy to observe during the spring and summer in most towns and cities in the eastern United States. Look for masses of small blackish brown ants struggling together on sidewalks or lawns. The combatants are members of rival colonies of the common pavement ant, *Tetramorium caespitum.* Thousands of individuals may be involved, and the battlefield typically occupies several square feet of the grassroots jungle.

Although some aggressive behavior in one form or another is characteristic of virtually all human societies (even the gentle !Kung Bushmen until recently had a murder rate comparable to that of Detroit and Houston), I know of no evidence that it constitutes a drive searching for an outlet. Certainly, the conduct of animals cannot be used as an argument for the widespread existence of such a drive.

In general, animals display a spectrum of possible actions, ranging from no response at all, through threats and feints, to all-out attack; and they select the action that best fits the circumstances of each particular threat. A rhesus monkey, for example, signals a peaceful intention toward another troop member by averting its gaze or approaching with conciliatory lip-smacking. A low level of hostility is conveyed by an alert, level stare. The hard look you receive from a rhesus when you enter a laboratory or the primate building of a zoo is not simple curiosity—it is a threat. From that point onward the monkey conveys increasing levels of confidence and readiness to fight by adding new components one by one or in combination: the mouth opens in an apparent expression of astonishment, the head bobs up and down, explosive *ho!*s are uttered, and the hands slap the ground. By the time the rhesus is performing all these displays, and perhaps taking little forward lunges as well, it is prepared to fight. The ritualized performance, which up to this point has served to demonstrate precisely the mood of the animal, may then give way to a shrieking, rough-and-tumble assault in which hands, feet, and teeth are used as weapons. Higher levels of aggression are not directed exclusively at other monkeys. Once, in the field, I had a large male monkey reach the hand-slapping stage three feet in front of me when I accidentally frightened an infant monkey that may or may not have been a part of the male's family. At that distance, the male looked like a small gorilla. My guide, Stuart Altmann of the University of Chicago, wisely advised me to avert my gaze and to look as much as possible like a subordinate monkey.

Despite the fact that many kinds of animals are capable of a rich, graduated repertoire of aggressive actions, and despite the fact that aggression is important in the organization of their societies, it is possible for individuals to go through a normal life, rearing offspring, with nothing more than occasional bouts of playfighting and exchanges of lesser hostile displays. The key is the environment: frequent intense display and escalated fighting are adaptive responses to certain kinds of social stress that a particular animal may or may not be fortunate enough to avoid during its lifetime. By the same token, we should not be surprised to find a few human cultures, such as the Hopi or the modern Aborigines of Australia, in which aggressive interactions are minimal. In a word, the evidence from comparative studies of animal behavior cannot be used to justify extreme forms of aggression, bloody drama, or violent competitive sports practiced by man.

This brings us to the topic that, in my experience, causes the most difficulty in discussions of human sociobiology: the relative importance of genetic versus environmental factors in the shaping of behavioral traits. The very notion that genes control behavior in human beings is scandalous to some scholars. They are quick to project a political scenario in which genetic determinism leads to support for the status quo and continued social injustice. Seldom do they entertain an equally plausible scenario, one in which complete cultural determinism leads to support for authoritarian mind control and worse injustice. Both sequences are highly unlikely unless politicians or ideologically committed scientists are allowed to dictate the uses of science. Then anything goes.

Concern over the implications of sociobiology usually proves to be due to a simple misunderstanding about the nature of heredity. Let me try to set the matter straight as briefly but fairly as possible. *What the genes prescribe is not necessarily a particular behavior but the capacity to develop certain behaviors and, more than that, the tendency to develop them in various specified environments.* Suppose that we could enumerate all conceivable behavior belonging to one category—say, all the possible kinds of aggressive responses—and for convenience label them by letters. In this imaginary example, there might be exactly twenty-three such responses, which we designate A through W. Human beings do not and cannot manifest all the behaviors; perhaps all societies in the world taken together employ A through P. Furthermore, they do not develop each of these with equal facility; there is a strong tendency under most possible conditions of child rearing for behaviors A through G to appear, and consequently H through P are encountered in very few cultures. It is this *pattern* of possibilities and probabilities that is inherited.

To make such a statement wholly meaningful, we must go on to compare human beings with other species. We note that hamadryas baboons can perhaps develop only F through J, with a strong bias toward F and G, while one kind of termite can show only A and another kind of termite only B. Which behavior a particular human being displays depends on the experience received within his or her own culture, but the total array of human possibilities, as opposed to baboon or termite possibilities, is inherited. It is the evolution of this pattern which sociobiology attempts to analyze.

We can be more specific about human patterns. It is possible to make a reasonable inference about the most primitive and general human social traits by combining two procedures. The first is to note the most widespread qualities of hunter-gatherer societies. Although the behavior of the people is complex and intelligent, the way of life to which their cultures are adapted is primitive. The human species evolved with such an elementary economy for hundreds of thousands of years; thus, its innate pattern of social responses can be expected to have been principally shaped by this way of life. The second procedure is to compare the most widespread hunter-gatherer qualities with similar behavior displayed by the species of langurs, colobus, macaques, baboons, chimpanzees, gibbons, and other Old World monkeys and apes that, together, constitute man's closest living relatives.

When we see the same pattern of traits occurring in all hunter-gatherer societies—and in most or all of the primates—we can conclude that it has been subject to relatively little evolution. Its possession by hunter-gatherers indicates (but does not prove) that the pattern was also possessed by man's immediate ancestors; the pattern also belongs to the class of behaviors least prone to change even in economically more advanced societies. On the other hand, when we see that a behavior varies a great deal among the primate species, we can infer that it is less resistant to change.

The list of basic human patterns that emerges from this screening technique is intriguing: (1) the number of intimate group members is variable but normally 100 or less; (2) some amount of aggressive and territorial behavior is basic, but its intensity is graduated, and its particular forms cannot be predicted with precision from one culture to another; (3) adult males are more aggressive and are dominant over females; (4) the societies are to a large extent organized around prolonged maternal care and extended relationships between mothers and children; and (5) play, including at least mild forms of contest and mock-aggression, is keenly pursued and probably essential to normal development.

We must then add the qualities that are so distinctively and ineluctably human that they can be safely classified as genetically based: the overwhelming drive of individuals to develop some form of a true, semantic language; the rigid avoidance of incest by taboo; and the weaker but still strong tendency for sexually bonded women and men to divide their labor into specialized tasks.

In hunter-gatherer societies, men hunt and women stay at home. This strong bias persists in most agricultural and industrial societies and, on that basis alone, appears to have a genetic origin. No solid evidence exists as to when the division of labor appeared in man's ancestors or how resistant to change it might be during the continuing revolution for women's rights. My own guess is that the genetic bias is intense enough to cause a substantial division of labor even in the most free and most egalitarian of future societies.

Substantial evidence exists that boys show persistently more mathematical and less verbal ability than girls on average and are more aggressive from the first hours of social play at age two to manhood. Thus, even with identical education and equal access to all professions, men seem likely to continue to play a disproportionate role in political life, business, and science. But such an outcome is only a guess and, even if correct, could not be used to argue for anything less than gender-blind admission and free personal choice.

Certainly, there are no a priori grounds for concluding that the males of a predatory species must be a specialized hunting class. Among chimpanzees, males are the hunters, a phenomenon that may be suggestive in view of the fact that these apes are by a wide margin our closest living relatives. But among lions, the females are the providers, typically working in groups with their cubs in tow. The stronger and largely parasitic males hold back from the chase but rush in to claim first share of the meat when the kill has been made. Wolves and African wild dogs follow still another pattern: adults of both sexes in these very aggressive species cooperate in the hunt.

There is a dangerous trap in sociobiology, one that can be avoided only by constant vigilance. The trap is the naturalistic fallacy of ethics, which uncritically concludes that what is, should be. The "what is" in human nature is to a large extent the heritage of a Pleistocene hunter-gatherer existence. The demonstration of any genetic bias cannot be used to justify a continuing practice in present and future societies. Since most of us live in a radically new environment of our own making, the pursuit of such a practice would be bad biology; and, like all bad biology, it would invite disaster. For example, the tendency under certain conditions to conduct warfare against competing groups might well be in our genes, having been advantageous to our Neolithic ancestors, but it could lead to global suicide now. To rear as many healthy children as possible was long the road to security; yet with the population of the world brimming over, such a strategy is now the way to environmental disaster.

Our primitive old genes will therefore have to carry the

load of much more cultural change in the future. To an extent not yet known, we trust—we insist—that human nature can adapt to more encompassing forms of altruism and social justice. Genetic biases can be trespassed, passions averted or redirected, and ethics altered; and the human genius for making contracts can continue to be applied to achieve healthier and freer societies. Yet the mind is not infinitely malleable. Human sociobiology should be pursued and its findings weighed as the best means we have of tracing the evolutionary history of the mind. In the difficult journey ahead, during which our ultimate guide must be our deepest and, at present, least understood feelings, surely we cannot afford an ignorance of history.

HUMANITY SEEN FROM A DISTANCE

All man's troubles arise from the fact
that we do not know what we are
and do not agree on what we want to be.

VERCORS (JEAN BRULLER),
You Shall Know Them (1953)

Here is the commencement address of the distinguished dean of the faculty of the International Termite University:

On one thing we can surely agree! We are the pinnacle of 3 billion years of evolution, unique by virtue of our high intelligence, employment of symbolic language, and diversity of cultures evolved over hundreds of generations. Our species alone has sufficient self-awareness to perceive history and the meaning of personal mortality. Having largely escaped the sovereignty of our genes, we now base social organization mostly or entirely upon culture. Our universities disseminate knowledge from the three great branches of learning: the natural sciences, the social sciences, and the termitities. Since our ancestors, the macrotermitine termites, achieved 10-kilogram weight and larger brains during their rapid evolution through the later Tertiary period and learned to write with pheromone script, termitistic scholarship has refined ethical philosophy. It is now possible to express the deontological imperatives of moral behavior with precision. These imperatives are mostly self-evident and universal. They are the very essence of termitity. They include the love of darkness and of the deep, saprophytic, basidiomycetic penetralia of the soil; the centrality of colony life amidst a richness of war and trade among colonies; the sanctity of the physiological caste system; the evil of personal reproduction by worker castes; the mystery of deep love for reproductive siblings, which turns to hatred the instant they mate; rejection of the evil of personal rights; the infinite aesthetic pleasures of pheromonal song; the aesthetic pleasure of eating from nestmates' anuses after the shedding of the skin; the joy of cannibalism and surrender of the body for consumption when sick or injured (it is more blessed to be eaten than to eat); and much more . . .

Some termitistically inclined scientists, particularly the ethologists and sociobiologists, argue that our social organization is shaped by our genes and that our ethical precepts simply reflect the peculiarities of termite evolution. They assert that ethical philosophy must take into account the structure of the termite brain and the evolutionary history of the species. Socialization is genetically channeled and some forms of it all but inevitable. This proposal has created a major academic controversy. Many scholars in the social sciences and termitities, refusing to believe that termite nature can be better understood by a study of fishes and baboons, have withdrawn behind the moat of philosophical dualism and reinforced the crenellated parapets of the formal refutation of the naturalistic fallacy. They consider the mind to be beyond the reach of materialistic biological research. A few take the extreme view that conditioning can alter termite culture and ethics in almost any direction desired. But the biologists respond that termite behavior can never be altered so far as to resemble that of, say, human beings. There is such a thing as a biologically based termite nature . . .

I have concocted this termitocentric fantasy to illustrate a generalization strangely difficult to explain by conventional means: that human beings possess a species-specific nature and morality, which occupy only a tiny section in the space of all possible social and moral conditions. If intelligent life exists on other planets (and the consensus of astronomers and biochemists is that it does, in abundance), we cannot expect it to be hominoid, mammalian, eucaryotic, or even DNA based. We should rescue the contemplation of other civilizations from science fiction. Real science tries to characterize not just the real world but all possible worlds. It identifies them within the much vaster space of all conceivable worlds studied by philosophers and mathematicians.

The social sciences and humanities have been blinkered by a steadfastly nondimensional and nontheoretical view of mankind. They focus on one point, the human species, without reference to the space of all possible species natures in which it is embedded. To be anthropocentric is to remain unaware of the limits of human nature, the significance of biological processes underlying human behavior, and the deeper meaning of long-term genetic evolution. That larger perspective can be gained only by moving back from the species, step by step, and taking a deliberately more distanced view.

In order to see the significance of multidimensionality, consider human social behaviors as a frequency-distribution function. The sociologist is perhaps closest of all to the array described by the function. Immersed in minute details of local culture, the typical sociologist fills the role of the local naturalist among the social scientists. He is not much concerned with the limits and ultimate meaning of human behavior. Indeed, he is likely to be oblivious to such distant matters, for the intricacy of detail seen in literate cultures is more than sufficiently important and absorbing to hold the attention of a first-rate scholar. The anthropologist and primatologist take a more distant view and are the equivalent of biogeographers. They have an interest in global patterns in the distribution of social traits, and they search for rules and laws to explain these peculiarities. The zoologist is the most removed. His concern is the tens of thousands of social species among the colonial invertebrates, social insects, and nonhuman vertebrates. The diversity he sees is enormous, but there is sufficient convergence in some categories of behavior among otherwise disparate taxonomic groups to raise in his mind the hope that general laws governing their genetic evolution might be adduced, just as studies of rats, fruit flies, and colon bacteria have yielded principles of genetics and physiology that could then be extended to human beings.

Of course, human social behavior has unique qualities unlikely to be predicted from a general, animal-based sociobiology. It cannot be compared to the purely mechanical behavior of human chromosomes and neuron membranes, which function almost exactly like those of rodents and insects. The human social repertoire now evolves along a dual track of inheritance: conventional genetic transmission, which is altered by conventional Darwinian natural selection; and cultural transmission, which is Lamarckian (traits acquired by the individual's adaptation are passed directly to offspring) and much swifter. Furthermore, unique features of organization exist: fully symbolic, endlessly productive language; long-remembered contracts based on convention; a complex materials-based culture; and religion. But the fact that humankind has entered a new zone of evolution is not evidence that the species has shed genetic constraint. Nor does sublimity necessarily elevate a species above biology. Traits that intelligent beings regard as transcendent may have arisen as biological adaptations while remaining obedient to genetic programs. The migratory flight of the golden plover from the Yukon to Patagonia and back is a marvel, but its brain and wings are made from organic polymers, and the 10,000-mile route of its journey is as necessary to the completion of its life cycle as its daily meal of beach fleas and insects. Substantial evidence exists that human behavior as a whole, including the most complex forms subject to the greatest cultural variation, is both genetically constrained and to some degree ultimately adaptive in the strict Darwinian sense. Thus social theory can be regarded as continuous with evolutionary biology.

If the perspective of the social sciences and humanities has been nondimensional in space, it has been equally restricted in time. This may seem a strange statement given that the examination of historical change is undeniably at the heart of each of the major disciplines. But all the analysis is based on a single species and, beyond that, on what is assumed to be a single genotype—the principle of the psychic unity of humankind. This conception of human sociality, though comforting, is inadequate for the needs of social theory. The evidence is strong that human populations vary to a degree typical of animal populations in behavioral traits, in particular in the genetic components of number ability, word fluency, memory, perceptual skill, psychomotor skill, extroversion–introversion, proneness to homosexuality, proneness to alcoholism, susceptibility to certain forms of neurosis and psychosis, timing of language acquisition and other major steps in cognitive development, age at first sexual activity, and other individual phenotypes that affect social organization. There is also evidence of geographic variation across human populations, in other words "racial" differences, in the earliest motor and temperament development of newborns.

Although genetic evolution is slow, it can occur rapidly enough to differ in rate from cultural evolution by only one or two orders of magnitude. Under only moderate selection pressures, one gene can be mostly substituted for another throughout an entire population in as few as ten generations, a period of only 200 or 300 years in the case of human beings. A single gene can profoundly alter behavior, especially when it affects the threshold of response or level of excitability. However, new, complex patterns of behavior are based on multiple genes, which can be assembled only over much longer periods, perhaps hundreds or even thousands of generations. For this reason we do not expect to find that human nature has been altered greatly during historical times, or that people in industrial societies differ basically from those in preliterate, hunter-gatherer societies. But the possibility that some genetic change has occurred has not been eliminated, and it cannot be assumed that small amounts of genetic change are easily washed out by the effects of socialization during the lifetimes of individuals.

If these elementary estimates are correct, significant elements of behavior might have originated within the past 100,000 years. In fact contemporary human nature need not be the product of the history of the ancestral *Australopithecus afarensis–Homo habilis* line 2 to 4 million years ago. It is more likely a biogram shaped gradually throughout the history of *Homo*, up to and including the historic period. Thus social theory could profit by extending its reach just beyond the historical period dominated by cultural evolution to the near prehistoric period during which more nearly balanced combinations of genetic and cultural change occurred.

CULTURE AS A BIOLOGICAL PRODUCT

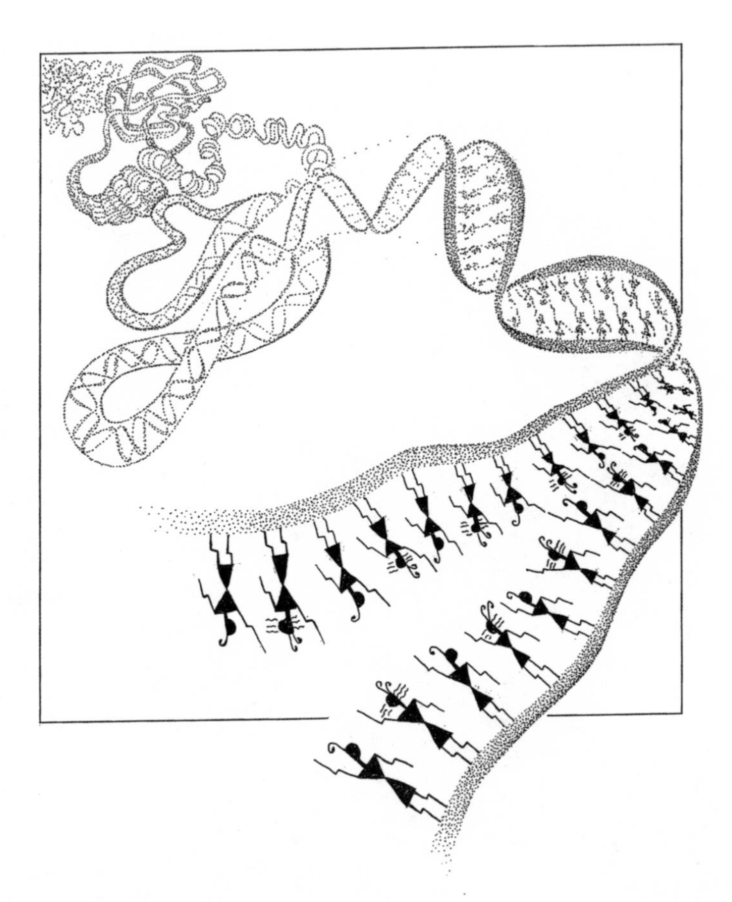

THIS IS THE ESSENCE OF THE matter as I understand it: culture is ultimately a biological product. As biology progresses as a science, it is sure to alter our understanding of social behavior and institutions. A large part of the variance in personality and cognition, in many cases half or more, is hereditary in origin. Even then the total amount due to heredity and environment combined is only a minute fraction of the amount conceivable, because cognitive development is severely constrained by genetically prescribed rules common to human beings. It has been said that there are no genes for building airplanes. That of course is true. But people build airplanes to conduct the primitive operations of human beings, including war, tribal reunions, and bartering, which conform transparently to their biological heritage. Culture conforms to an important principle of evolutionary biology: most change occurs to maintain the organism in its steady state.

The principal driving force of genetic evolution of all organisms studied to date is natural selection, the differential contribution to the next generation by various genetic types belonging to the same population. This is the process often called Darwinism, to distinguish it from mutation pressure, orthogenesis (straight-line evolution), and other conceivable driving forces. A great deal of evolution at the level of molecular structure appears to be due to genetic drift, the random substitution of alleles affecting amino acid substitutions in proteins. Nevertheless, the main features of anatomy, physiology, and behavior are ultimately ascribable to natural selection.

A differential contribution to the next generation can be achieved by the interplay of two advantages gained: longer life and greater reproduction. Individuals can put more of their genes into the future by reproducing as fast as possible, counting on at least a few surviving to maturity. This is the *r* strategy of reproduction. Or they can get the same result by producing only a small number of high-quality offspring and nurturing them carefully to be sure that most or all reach maturity in good condition. This is the *K* strategy of reproduction. Which of the two strategies works best depends on the environment. When resources are unpredictable, with a good chance of extinction from one place or time to another, the *r* strategy works best. When resources are dependable and fixed, so that land tenure is important, the *K* strategy is more likely to succeed. Biologists often place species and genetic strains along an *r*–*K* continuum, relating their reproductive strategies to the environment in which the species and strains have evolved. It is also possible to have genotypes that program switches from one strategy to the next as conditions change. Human beings occupy a small segment of the r–K continuum near the K extreme.

Gene-Culture Coevolution

Human evolution is a unique dual-track system compounded of genetic change and cultural change. On the one hand genetic change brought about an extremely rapid growth of the human brain, with a 3.2-fold increase in the volume of the cerebral cortex alone from the time of *Homo habilis* 2 million years ago to the appearance of early *Homo sapiens* about 500,000 years ago, accompanied by profound architectural innovations in the larynx and speech centers of the brain. Cultural change is much faster, but it is limited and directed by the restricting properties of the brain and sensory apparatus.

Most of the difficulty in human sociobiology arises not from the differences in procedure and language between biologists and social scientists, although these are real enough, but from the fact that the subject of common interest, the interaction of biological and cultural evolution, remains largely unexplored. We all know that human social behavior is transmitted through learning and culture. We also know that the distinctive properties of cognition, ranging from a sensory perception to memory and decision making, have a powerful effect on culture. Culture is determined ultimately by the mental development of individual human beings. The properties of this development can be characterized by epigenetic rules, which are defined as any regularities that bias behavior in a particular direction. To give a quick example, human beings are highly audiovisual and depend very little on smell and taste in comparison with the great majority of animal species. This biological property redounds to a much richer vocabulary describing hearing and vision than is the case for smell and taste. In various languages around the world, about two-thirds to three-fourths of all the words applying to the senses describe hearing and vision, while one-tenth or fewer describe smell and taste.

Genetic evolution thus affects cultural evolution. Conversely, cultural evolution affects biological evolution, by creating the environment in which the genes (the ones prescribing epigenetic rules) are tested through natural selection. Genes and culture are in fact inseverably linked. Changes in one inevitably force changes in the other, resulting in what has come to be called gene-culture coevolution. The process is believed to occur as follows:

- The genes prescribe the rules of development (the epigenetic rules) by which the individual mind is assembled.
- The mind grows by absorbing parts of the culture already in existence.

- The culture is created anew in each generation by the summed decisions and innovations of all members of the society.
- Some individuals possess epigenetic rules enabling them to survive and reproduce better in the contemporary culture than other individuals. This genetic fitness can be enhanced either by direct selection, the furthering of direct descendants, or by kin selection, the furthering of collateral kin in addition to direct descendants.
- The more successful epigenetic rules spread through the population, along with the genes that encode them. Put another way, the population evolves genetically with reference to the epigenetic rules.

To summarize to this point, culture is created and shaped by biological processes while the biological processes are simultaneously altered in response to cultural change. This process is not difficult to envisage, but the rates at which the two forms of evolution occur and the tightness of the linkages between them remain largely unsolved problems.

The Units of Culture

The principal theoretical difficulties of the social sciences are two in number. First, in the study of culture there are no "natural kinds," basic atomic units equivalent to genes, cells, and organisms that can form the base of permutational operations in analysis. The lack of natural kinds guarantees the second difficulty, "nomic isolation." Each major discipline—anthropology, sociology, political science, and so forth—has been required to develop its own conceptual base and language.

The discovery of natural kinds in culture would represent a key theoretical advance in the social sciences. Most scholars appear to believe that such units either do not exist or, if they do exist, cannot be derived by any means currently available. However, there is some reason to believe that natural units do exist and are built on the natural units of semantic memory. Semantic memory comprises words and symbolic manipulation, as opposed to episodic memory, which entails running sequences of visual and other sensory experience. It tends to organize impressions into discrete clusters. Experimental studies have revealed that the cuts are made around objects or abstractions that have the most attributes in common. Hence although categories such as "tree," "dog," and "house" do not exist in the real world, they are collections of objects that share a relatively large number of stimuli most easily processed by the brain. Children move easily into this mode of memory formation, performing equally well on objects or collections of objects. They organize certain identifying stimuli into ensembles (such as "cookies" versus "cakes" and "chairs" versus "stools") that are almost as sharply demarcated as the separate objects themselves.

The brain speeds processing still further by compounding the clusters hierarchically into larger assemblages that possess a discrete, interchangeable form. The units of semantic memory, which are experienced as objects or abstractions, are appropriately called nodes, thus aligning the description to the nodes and the links between nodes envisioned in spreading-activation models of memory storage and recall. There are at least three levels of nodes. Concepts, the most elementary clusters, are signaled by words or phrases (such as "dogs" and "hunt"). Propositions are signaled by phrases, clauses, or sentences expressing objects and relationships ("dogs hunt"). Finally, schemata are signaled by sentences and larger units of text (the "technique of hunting with dogs"). Node-link structures were originally proposed by psychologists as theoretical representations, but they have gained considerable substance through methods that detect their organization. Node-link structures steadily enlarge in size and complexity in the growing child, and the main steps in the growth correspond at least roughly to the Piagetian stages of mental development. The stages are not mere accidents of personal growth but general processes that show some regularity across cultures. Hence, in a manner important for the entire relation of biology to culture, the semantic mechanisms of culture formation are more robust and consistent than the final products they generate.

For each concept the brain tends to select a prototype that constitutes the standard, such as a particular wavelength and intensity to form the idealized color red or a particular body shape and size to form the typical "dog." Given an array of similar variants, the mind can deduce a standard near the average of the variants and use it as the prototype even without having perceived any example of it directly. The most important result for gene-culture coevolution is that the divisions are created and labeled, even when the stimuli being processed vary continuously. In short, the mind automatically imposes a semidiscrete, hierarchical order upon the world.

Most of the concepts that make up the basic units of semantic memory are subject to purely phenotypic variation arising from the particularities of cultural history. Nevertheless there is a tendency for those belonging to at least a few categories to occur consistently across cultures. As Eleanor Rosch has shown, such categories include elementary geometric forms (square, circle, equilateral triangle), the facial expressions of six basic emotions (happiness, sadness, anger, fear, surprise, disgust), and the basic colors (red, yellow, green, blue).

The level of the node of semantic memory, whether concept (the most elementary recognizable unit), proposition, or schema, determines the complexity of the generated behavior or artifact maintained in the culture. For example, the differentiation of letters or ideographs is at the level of the concept, the initial verbal reaction to a stranger is a proposition, and the expression of an incest taboo is a schema. If this model of semantic memory holds, new discoveries refining the hierarchy of memory nodes can be expected to advance the identification of culture units, or "culturgens," in the same manner that advances in cellular chemistry have improved our comprehension of the gene and studies of population structure have refined our understanding of biological species.

Although a direct correspondence between nodes and generative units of culture appears feasible at lower levels of organization, there is no reason to expect the more complex constructions of culture to be mapped onto semantic nodes in a one-to-one fashion. Marriage ceremonies and temple architecture, for example, are the outcomes of numerous interlocking behaviors that result from cognitive activity with multiple culturgens. These in turn vary according to the particularities of local history. Nevertheless, each can be realistically interpreted as the outcome of cognitive development, which is attained principally through the assembly of node-link networks. Cultural evolution is the shifting of the outer phenotypes of behavior and artifacts through the insertion and combination of their basic generative structures in semantic memory. The compound structures of culture arise from the semantic nodes.

Epigenetic Rules

The epigenetic rules of cognitive development determine the manner in which the nodes are created and combined to form the semantic networks—and hence, culture.

These physiological processes impose a strict filtering of stimuli from the environment and alter each step of cognition thereafter, from short-term memory and storage in long-term memory to recall, feeling, reveries, and decision making.

The most fully analyzed example of the biological channeling of culture by the processes of filtering and biasing arises in the vocabulary of vision. Light intensity is perceived as a continuum; if the light in a room is raised or lowered gradually with a dimmer switch, the conscious brain perceives the change as a continuous progression along a more or less even gradient. There are no steps or benchmarks, and consequently languages contain relatively few words to describe the variation of light intensity. In contrast, normally sighted individuals see variation in wavelength not as a continuously varying property of light (which it is) but as the four basic colors of blue, green, yellow, and red, along with various blends in the intermediate zones. If a room is flooded with monochrome light of short wavelength (blue), and the wavelength is then gradually increased, the change is seen as a series of steps from one basic color to another. The physiological basis of this illusion is partially known. The innate human color classification starts from the differentiation of the retinal cones into three types, whose maximal sensitivities correspond to blue, green, and red. The light-sensitive pigments in the cones are membrane proteins, with retinal, a pigment molecule, attached in each case to an apoprotein. When the retinal is altered by a photon of light to pass from the *cis* to the *trans* state, the apoprotein is induced into a configurational change which in turn depolarizes an afferent nerve cell. The red and green pigments have recently been identified, and the genes specifying them located and sequenced. The Mendelian genetics of color blindness has also been partially worked out. Further encoding of color occurs in four classes of interneurons in the lateral geniculate nuclei of the thalamus that lead to the processing centers of the visual cortex.

How do such facts bear on culture? The epigenetic constraints in color perception are reflected in languages of all cultures thus far examined. In a classic study by Brent Berlin and Paul Kay, native speakers of twenty languages around the world (including Arabic, Bulgarian, Cantonese, Catalan, Hebrew, Ibidio, Japanese, Thai, Tzeltan, Urdu, and others) were shown arrays of chips classified by color and brightness in the Munsell system. They were asked to place each of the principal color terms of their language within this two-dimensional array. The results show clearly that the languages have evolved in a way that conforms closely to the epigenetic rules of color discrimination. The words fall into largely discrete clusters that correspond, at least approximately, to the principal colors that are innately distinguished.

The intensity of learning bias was further revealed by another experiment conducted by Eleanor Rosch. In looking for "natural categories" of cognition, Rosch exploited the fact that the Dani people of New Guinea have no words to denote color; they speak only of *mili* (roughly, "dark") and *mola* ("light"). Rosch addressed the following question: if Dani adults set out to learn a color vocabulary, would they do so more readily if the color terms corresponded to the principal innate hues? In other words, would cultural innovation be channeled to some extent by the innate genetic constraints? Rosch divided 68 volunteer Dani into two groups. She taught one a series of newly invented color terms placed on the principal hue categories of the array (blue, green, yellow, red), where most of the natural vocabularies of other cultures are located. She taught a second group a series of new terms placed off center, away from the main clusters formed by other languages. The first group of volunteers, following the "natural" propensities of color perception, learned about twice as quickly as those given the competing, less natural color terms. They also selected these terms more readily when given a choice.

In a parallel experiment on "psychoaesthetics," Gerda Smets measured the degree of physiological arousal in adults caused by geometric designs of varying degrees of complexity. The measure she used was alpha wave blockage, generally interpreted to be an index of arousal even when unaccompanied by conscious awareness. A sharp peak of maximum response was obtained with computer-generated figures at 20 percent redundancy, the amount found, for example, in a maze with between 10 and 20 angles. Less redundancy and more redundancy were far less stimulating. It does not seem coincidental that 20 percent is approximately the amount of complexity in logotypes, ideographs, frieze design, grille work, and other designs chosen for instant recognition and aesthetic pleasure. In other words, the development of art and written language may be strongly influenced by an innate constraint in cognition.

The innate bias in learning, described by psychologists as "prepared" (bias toward) and "counterprepared" (bias against), is perhaps most strikingly illustrated by phobias. These are extreme, irrational fears associated with nausea, cold sweat, and other reactions of the central nervous system. It is notable that the phobias are most easily evoked by many of the greatest dangers of mankind's ancient environment, including tight spaces, heights, thunderstorms, running water, snakes, and spiders, but are rarely evoked by the greatest dangers of modern technological society, including guns, knives, automobiles, explosives, and electric sockets.

The Translation from Genes to Culture

In order to picture gene-culture coevolution more clearly, imagine two alien civilizations on distant planets. Both have about the same level of cultural sophistication as human beings, and both transmit virtually all of their cultures by means of learning. However, in one of the civilizations only a single version of each category of learning can be transmitted: one language, one love song, one marriage ceremony, one mode of warfare, and so forth. In this extreme form, a "pure genetic transmission" of the culture, the genes restrict the learning process—even though the culture is taught in classrooms, recorded in books, and so forth. The conception is not too far-fetched. Individuals of this species are like the white-crowned sparrows of California, which must hear the song of their own species in order to learn it but are impervious to all other songs.

The second alien species outwardly resembles the first, but it possesses a totally blank-state mind. All cultural possibilities are open to the inhabitants. They can be taught any language, any song, any martial tactic with approximately equal ease. In this "pure cultural transmission" scenario, the genes direct the construction of the body and brain but not the behavior. The mind is entirely a product of the accidents of history, including the place the aliens live in, the foods they encounter, and the stray inventions of words and gestures.

Human beings are of course in between these extremes. Our social behavior is based on gene-culture transmission: an immense array of possibilities can be learned, innovation occurs frequently, but biological properties in the sense organs and brain make it more likely that certain choices will be preferred or at least more easily learned than others. In some categories, such as incest avoidance, the choices are narrowly constrained. In others, such as the semantic content of particular languages (but not the deep grammatical properties), the choices are very broad and more nearly equipotent.

This conception of mental development brings us to the question of variation in the choice of culturgens among members of a given society and among entire societies. The evolution of culture displays some striking parallels with genetic evolution. Innovations appear in the population in the manner of mutations, spread like genes, and are favored or abandoned by processes resembling natural selection and random drift. The interaction of these biologically based entities with the environment is at least as complex and analytically formidable as that controlling conventional genetic evolution. Among the variables that must eventually be taken into account are the particular environment in which the society lives, the degree of its contact with surrounding cultures, the accidents of history, and the genetic variation among its members.

Using a separate language, social scientists and humanists have already explored these matters in considerable depth. But although their accounts of cultural variation are rich and illuminating, they do not penetrate to the biological underpinnings of mental life. Ordinary inductive descriptions of behavior and culture can in fact never achieve this end; it is just not enough, as Darwin said, to attack the citadel itself. The more promising approach is less direct. It reconstitutes cultural variation along with central tendencies in a combined analytic-synthetic fashion, from the bottom up, using the facts of biology and cognitive psychology to work into more complex social phenomena.

It is logical to begin such an analysis with the simple case of a human population that is genetically uniform with reference to the processes of cognition. In studies conducted in 1980–1982 and described in our 1984 book, *Promethean Fire*, Charles Lumsden and I chose as a goal the transition from individual learning and decision making to cultural diversity in the absence of genetic variation and in a relatively uniform environment. We set out to find which patterns of cultural diversity can be expected to result from different forms and degrees of bias in cognitive development, and we asked whether the observed patterns of cultural diversity were consistent with what is understood about this development.

We began with the simple observation that each individual comes to favor certain marital customs, modes of dress, ethical precepts, and so forth, from among those available. And every time individuals modify their memories or face decisions in everyday life, they enact intricate sequences of events in cognition that obey the peculiar, constraining properties of semantic memory. Not all the culturgens being processed are treated equally; cognition has not evolved as a wholly neutral filter, and the mind incorporates and uses certain culturgens more readily than others. Furthermore, the biases often shift with age, creating patterns that change with the demographic properties of societies.

Because such usage biases are both discrete and episodic in operation, they can be approximated by transition probabilities, which can then be converted into rates of change treated as Markov processes. Experience from sociological studies has shown that such models can incorporate memory and social context to an extent sufficient to fit real (but by no means all) data on choices made by individuals. We have examined ways of incorporating experience and memory to make the needed jump to cultural variation still more realistic.

In particular, the transition rates from one alternative choice to another are affected by the choices already made by others, in other words the cultural context. Few quantitative studies have been made of this social influence, but enough is known to establish that it varies substantially from one behavioral category to another. For example, sibling incest is avoided by individuals throughout their lives regardless of the preferences of others, whereas the direction of attention of individuals in street crowds rises steadily in conformity with others as the percentage looking in one direction increases.

With the aid of these mathematical techniques, it is

possible to translate decision making and the effects of social networks into patterns of cultural diversity. Although this phase of the work is theoretical, it has yielded several general results that are interesting enough to merit closer attention. First, the procedure identifies the quantitative description of cultural diversity most readily aligned with studies of cognitive science. This is the *ethnographic distribution*, comprising the relative frequencies of societies in which different percentages of the members use or at least prefer to use each of the competing culturgens. A simple ethnographic distribution would be the following: in 52 percent of the societies all members prefer outbreeding to incest, in 46 percent of the societies 99 percent prefer outbreeding, and in 2 percent of the societies 98 percent prefer it.

A notable finding of the models is that very substantial cultural diversity can be expected even when all of the societies are genetically biased to a rather high degree with respect to that particular category of cognition and behavior. Even though all of humanity may be genetically very likely to choose outbreeding over incest, substantial variation will still arise among the societies in the percentages of members choosing avoidance over acceptance. Because the mind is probabilistic in operation, what emerges is not a fixed percentage of individuals making one choice across all societies but rather the pattern of diversity, in other words, the form of the ethnographic distribution. A distinct ethographic curve will arise from each different degree of bias toward one culturgen and each different degree of sensitivity toward the choice already made by others in the society. For each category of cognition and behavior, human beings appear to have a distinctive degree of developmental bias and sensitivity. As a result the amount and pattern of cultural diversity can be expected to differ among these categories.

It is often argued that the existence of cultural diversity shows that there is no underlying genetic constraint. That conclusion, which at first may seem common sense, is incorrect: the mere occurrence of the diversity says nothing one way or the other about constraints. On the other hand, patterns of diversity can tell us a great deal. Another common misconception is that the existence of biological influence on diversity implies genetic differences between the societies. But as Lumsden and I have shown, diversity arises in distinctive patterns even in genetically uniform populations.

The models lead to another substantive result of the gene-to-culture theory. Quite small differences in bias and sensitivity of the magnitude demonstrated among different categories of human cognition and behavior are enough to generate strong differences among their patterns of cultural diversity. Most strikingly, the distributions pass from a single mode to multiple modes (a mode is a frequency higher than surrounding frequencies) rather rapidly as sensitivity is altered. These differences are great enough to be detected even with relatively crude ethnographic or sociological data. They show how studies of cognitive and social psychology can be fed directly into the data of anthropology and sociology as part of a general quantitative theory of culture.

Culture is deeply rooted in biology. Its evolution is channeled by the epigenetic rules of mental development, which in turn are genetically prescribed. We can envisage the full chain of causation from genetic prescription to the formation of culture and back again through natural selection to changes in gene frequencies. Gene-culture coevolution, as the reciprocating process is known, has been documented through part of the cycle, and some of the key steps have been addressed with analytic models. Its further exploration seems very promising for the future study of culture.

THE BIRD OF PARADISE: THE HUNTER AND THE POET

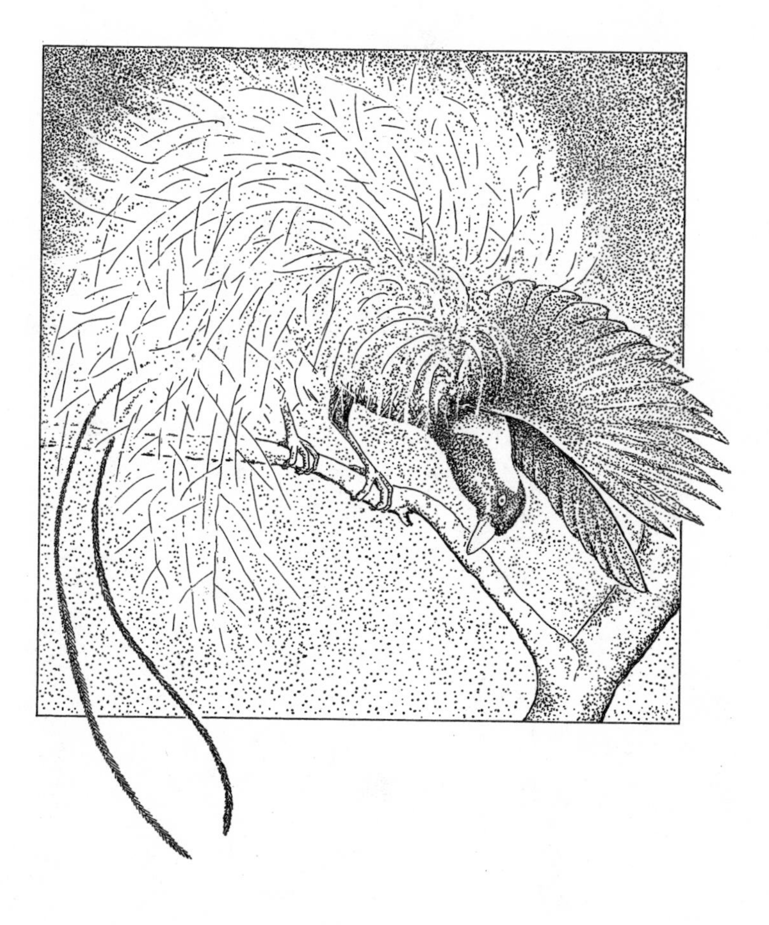

THE ROLE OF SCIENCE, LIKE that of art, is to blend proximate imagery with more distant meaning, the parts we already understand with those given as new into larger patterns that are coherent enough to be acceptable as truth. Biologists know this relation by intuition during the course of fieldwork, as they struggle to make order out of the infinitely varying patterns of nature.

Picture the Huon Peninsula of New Guinea, about the size and shape of Rhode Island, a weathered horn projecting from the northeastern coast of the main island. When I was twenty-five, with a fresh Ph.D. from Harvard and dreams of physical adventure in far-off places with unpronounceable names, I gathered all the courage I had and made a difficult and uncertain trek directly across the peninsular base. My aim was to collect a sample of ants and a few other kinds of small animals up from the lowlands to the highest part of the mountains. To the best of my knowledge I was the first biologist to take this particular route. I knew that almost everything I found would be worth recording, and all the specimens collected would be welcomed into museums.

Three days' walk from a mission station near the southern Lae coast brought me to the spine of the Sarawaget range, 12,000 feet above sea level. I was above treeline, in a grassland sprinkled with cycads, squat gymnospermous plants that resemble stunted palm trees and date from the Mesozoic era; closely similar ancestral forms might have been browsed by dinosaurs 80 million years before. On a chill morning when the clouds lifted and the sun shone brightly, my Papuan guides stopped hunting alpine wallabies with dogs and arrows, I stopped putting beetles and frogs into bottles of alcohol, and together we scanned the rare panoramic view. To the north we could make out the Bismarck Sea, to the south the Markham Valley and the more distant Herzog Mountains. The primary forest covering most of this mountainous country was broken into bands of different vegetation according to elevation. The zone just below us was the cloud forest, a labyrinth of interlocking trunks and branches blanketed by a thick layer of moss, orchids, and other epiphytes that ran unbroken off the tree trunks and across the ground. To follow game trails across this high country was like crawling through a dimly illuminated cave lined with a spongy green carpet.

A thousand feet below, the vegetation opened up a bit and assumed the appearance of typical lowland rain forest, except that the trees were denser and smaller and only a few flared out into a circle of blade-thin buttresses at the base. This is the zone botanists call the mid-mountain forest. It is an enchanted world of thousands of species of birds, frogs, insects, flowering plants, and other organisms, many found nowhere else. Together they form one of the richest and most nearly pure segments of the Papuan flora and fauna. To visit the mid-mountain forest is to see life as it existed before the coming of man thousands of years ago.

The jewel of the setting is the male Emperor of Germany bird of paradise *(Paradisaea guilielmi)*, arguably the most beautiful bird in the world, certainly one of the twenty or so most striking in appearance. By moving quietly along secondary trails you might glimpse one on a lichen-encrusted branch near the treetops. Its head is shaped like that of a crow—no surprise, since the birds of paradise and crows have a close common lineage—but there the outward resemblance to any ordinary bird ends. The crown and upper breast of the bird are metallic oil-green and shine in the sunlight. The back is glossy yellow, the wings and tail deep maroon. Tufts of ivory-white plumes sprout from the flanks and sides of the breast, turning lacy in texture toward the tips. The plume rectrices continue on as wirelike appendages past the breast and tail for a distance equal to the full length of the bird. The bill is blue-gray, the eyes clear amber, the claws brown and black.

In the mating season the male joins others in leks, common courtship arenas in the upper tree branches, where they display their dazzling ornaments to the more somberly caparisoned females. The male spreads his wings and vibrates them while lifting the gossamer flank plumes. He calls loudly with bubbling and flutelike notes and turns upside down on the perch, spreading wings and tail and pointing his rectrices skyward. The dance reaches a climax as he fluffs up the green breast feathers and opens out the flank plumes until they form a brilliant white circle around his body, with only the head, tail, and wings projecting beyond. The male sways gently from side to side, causing the plumes to wave gracefully as if caught in an errant breeze. Seen from a distance, his body now resembles a spinning and slightly out-of-focus white disk.

This improbable spectacle in the Huon forest has been fashioned by thousands of generations of natural selection in which males competed and females made choices, and the accouterments of display were driven to a visual extreme. But this is only one trait, seen in physiological time and thought about at a single level of causation. Beneath its plumed surface, the Emperor of Germany bird of paradise possesses an architecture marking the culmination of an equally ancient history, with details exceeding those that can be imagined from the elaborate visible display of color and dance.

Consider one such bird analytically, as an object of biological research. Encoded within its chromosomes is the developmental program that has led to a male *Paradisaea guilielmi*. Its nervous system is a structure of fiber tracts more complex than that of any existing computer, and as challenging as all the rain forests of New Guinea surveyed on foot. Someday microscopic studies will permit us to trace the events culminating in the electric commands carried by the efferent neurons to the skeletal-muscular system and to reproduce, in part, the dance of the courting male. We will be able to dissect and understand this machinery at the level of the cell through enzymatic catalysis, microfilament configuration, and active sodium transport during electric discharge. Because biology sweeps the full range of space and time, more and more discoveries will renew our sense of wonder at each step of research. Altering the scale of perception to the micrometer and millisecond, the cellular biologist's trek parallels that of the naturalist across the land. He looks out from his own version of the mountain crest. His spirit of adventure, as well as personal history of hardship, misdirection, and triumph, is fundamentally the same.

Described this way, the bird of paradise may seem to have been turned into a metaphor of what humanists dislike most about science: that it reduces nature and is insensitive to art, that scientists are conquistadors who melt down the Inca gold. But science is not just analytic; it is also synthetic. It uses artlike intuition and imagery. True, in the early analytic stages, individual behavior can be mechanically reduced to the level of genes and neurosensory cells. But in the synthetic phase even the most elementary activity of these biological units is seen to create rich and subtle patterns at the levels of organism and society. The outer qualities of *Paradisaea guilielmi*, its plumes, dance, and daily life, are functional traits open to a deeper understanding through the exact description of their constituent parts. They can be redefined as holistic properties that alter our perception and emotion in surprising ways.

There will come a time when the bird of paradise is reconstituted through a synthesis of all the hard-won analytic information. The mind, exercising a newfound power, will journey back to the familiar world of seconds and centimeters, where once again the glittering plumage takes form and is viewed at a distance through a network of leaves and mist. Once again we see the bright eye open, the head swivel, the wings extend. But the familiar motions are now viewed across a far greater range of cause and effect. The species is understood more completely; misleading illusions have given way to more comprehensive light and wisdom. With the completion of one full cycle of intellect, the scientist's search for the true material nature of the species is partially replaced by the more enduring responses of the hunter and poet.

What are these ancient responses? The full answer is available only through a combined idiom of science and the humanities, whereby the investigation turns back into itself. The human being, like the bird of paradise, awaits our examination in the analytic-synthetic manner. Feeling and myth can be viewed at a distance through physiological time, idiosyncratically, in the manner of traditional art. But they can also be penetrated more deeply than was ever possible in the prescientific age, to their physical basis in the processes of mental development, the brain structure, and indeed the genes themselves. It may even be possible to trace them back beyond the formation of cultures to the evolutionary origins of human nature. As each new phase of synthesis emerges from biological inquiry, the humanities will expand their reach and capability. In symmetric fashion, with each redirection of the humanities, science will add dimensions to human biology.

NATURE'S

ABUNDANCE

THE LITTLE THINGS THAT RUN THE WORLD

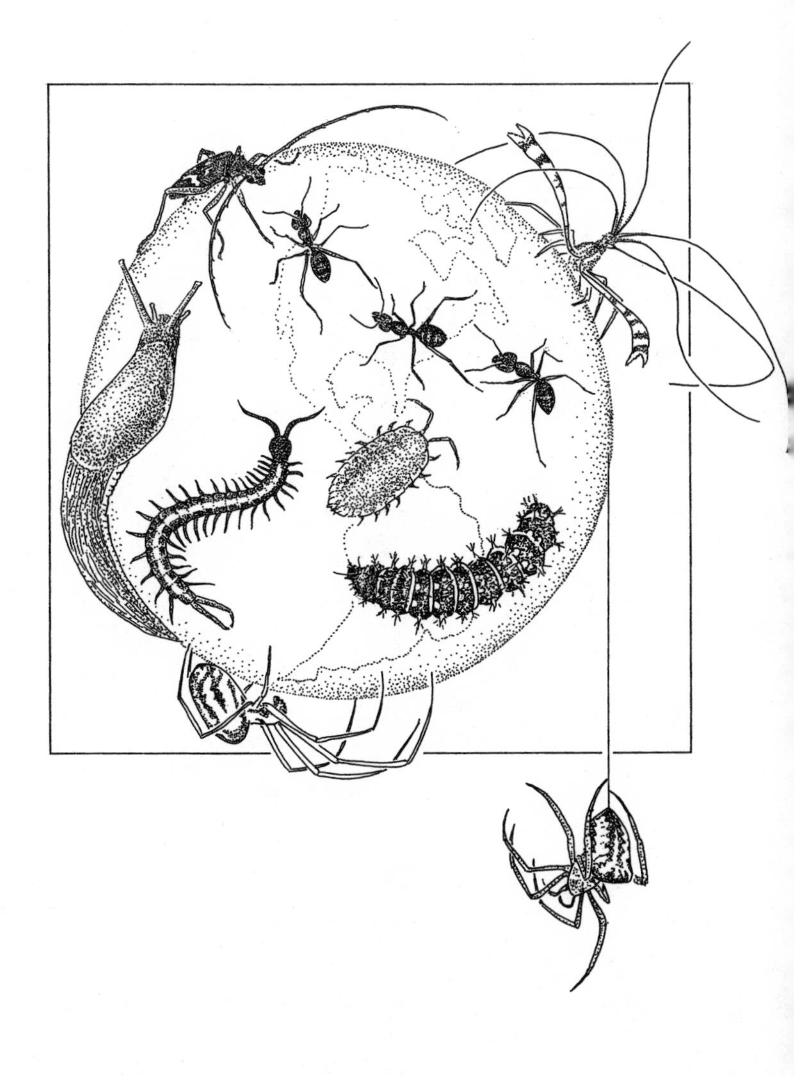

THERE ARE VASTLY MORE kinds of invertebrates than of vertebrates. In 1988, on the basis of a tabulation of the literature compiled with the help of specialists, I estimated that a total of 42,580 vertebrate species have been scientifically described, of which 6,300 are reptiles, 9,040 are birds, and 4,000 are mammals. In contrast, 990,000 species of invertebrates have been described, of which 290,000 alone are beetles—seven times the number of all the vertebrates together. Recent estimates have placed the number of invertebrate species on the Earth as high as 10 million, and possibly more.

We don't know with certainty why invertebrates are so diverse, but a commonly held opinion is that the key trait is their small size. Their niches are correspondingly small, and they can therefore divide up the environment into many more little domains where specialists can coexist. Among my favorite examples of such specialists living in microniches are the mites that live on the bodies of army ants: one kind is found only on the mandibles of the soldier caste, where it sits and feeds from the mouth of its host; another kind is found only on the hind foot of the soldier caste, where it sucks blood for a living; and so on through various bizarre configurations.

Another possible cause of invertebrate diversity is the greater antiquity of these little animals, which has given them more time to explore the environment. The first invertebrates appeared well back into Precambrian times, at least 600 million years ago. Most invertebrate phyla were flourishing before the vertebrates arrived on the scene, some 500 million years ago.

Invertebrates also rule the earth by virtue of sheer body mass. For example, in tropical rain forest near Manaus, in the Brazilian Amazon, each hectare (or 2.5 acres) contains a few dozen birds and mammals but well over one billion invertebrates, of which the vast majority are mites and springtails. There are about 200 kilograms dry weight of animal tissue in a hectare, of which 93 percent consists of invertebrates. The ants and termites alone compose one-third of this biomass. So when you walk through a tropical forest, or most other terrestrial habitats for that matter, or snorkel above a coral reef or some other marine or aquatic environment, vertebrates may catch your eye most of the time—biologists would say that your search image is for large animals—but you are visiting a primarily invertebrate world.

It is a common misconception that vertebrates are the movers and shakers of the world, tearing the vegetation down, cutting paths through the forest, and consuming most of the energy. That may be true in a few ecosystems such as the grasslands of Africa with their great herds of herbivorous mammals. It has certainly become true in the past few centuries in the case of our own species, which now appropriates in one form or other as much as 40 percent of the solar energy captured by plants. That circumstance is what makes us so dangerous to the fragile environment of the world. But in most parts of the world it is the invertebrates rather than the nonhuman vertebrates that are the movers and shakers. In Central and South America, for example, the leafcutter ants, rather than deer, or rodents, or birds, are the principal consumers of vegetation. A single colony of leafcutters contains several million workers. Columns of foragers travel 100 meters or more in all directions to cut forest leaves, flower parts, and succulent stems. Each day a typical mature colony collects about 50 kilograms of this fresh vegetation, more than the average cow. The workers excavate vertical galleries and living chambers as deep as 5 meters into the soil. The leafcutters and other kinds of ants, together with bacteria, fungi, termites, and mites, process most of the dead vegetation and return its nutrients to the plants to keep the great tropical forests alive.

Much the same situation exists in other parts of the world. The coral reefs are built out of the bodies of coelenterates. The most abundant animals of the open sea are copepods, tiny crustaceans forming part of the plankton. The mud of the deep sea is home to a vast array of mollusks, crustaceans, and other small creatures that subsist on the fragments of wood and dead animals drifting down from the lighted areas above, and on each other.

The truth is that we need invertebrates but they don't need us. If human beings were to disappear tomorrow, the world would go on with little change. Gaia, the totality of life on Earth, would set about healing itself and return to the rich environmental states of 100,000 years ago. But if invertebrates were to disappear, it is unlikely that the human species could last more than a few months. Most of the fishes, amphibians, birds, and mammals would crash to extinction about the same time. Next would go the bulk of the flowering plants and with them the physical structure of the majority of the forests and other terrestrial habitats of the world. The soil would rot. As dead vegetation piled up and dried out, narrowing and closing the channels of the nutrient cycles, other complex forms of vegetation would die off, and with them the last remnants of the vertebrates. The remaining fungi, after enjoying a population explosion of stupendous proportions, would also perish. Within a few decades the world would return to the state of a billion years ago, composed primarily of bacteria, algae, and a few other very simple multicellular plants.

In addition to these functions that make us completely dependent upon them, these little creatures that run the world provide us with an endless source of scientific exploration and naturalistic wonder. When you scoop up a double handful of soil almost anywhere except in the most barren deserts, you will find thousands of invertebrate animals, ranging in size from clearly visible to microscopic, from ants and springtails to tardigrades and rotifers. The biology of most of the species you hold is unknown: we have only the vaguest idea of what they eat, what eats them, and the details of their life cycle, and probably nothing at all about their biochemistry and genetics. Some of the species probably even lack scientific names. We have little concept of how important any of them are to our existence. Their study would certainly teach us new principles of science to the benefit of humanity. Each one is fascinating in its own right. If human beings were not so impressed by size alone, they would consider an ant more wonderful than a rhinoceros.

New emphasis should be placed on the conservation of invertebrates. Their staggering abundance and diversity should not lead us to think that they are indestructible. On the contrary, their species are just as vulnerable to extinction through human interference as are those of birds and mammals. When a valley in Peru or an island in the Pacific is stripped of the last of its native vegetation, the result is likely to be the extinction of several kinds of birds and some dozens of plant species. Whereas we are painfully aware of that tragedy, we fail to perceive that hundreds of invertebrate species will also vanish.

SYSTEMATICS ASCENDING

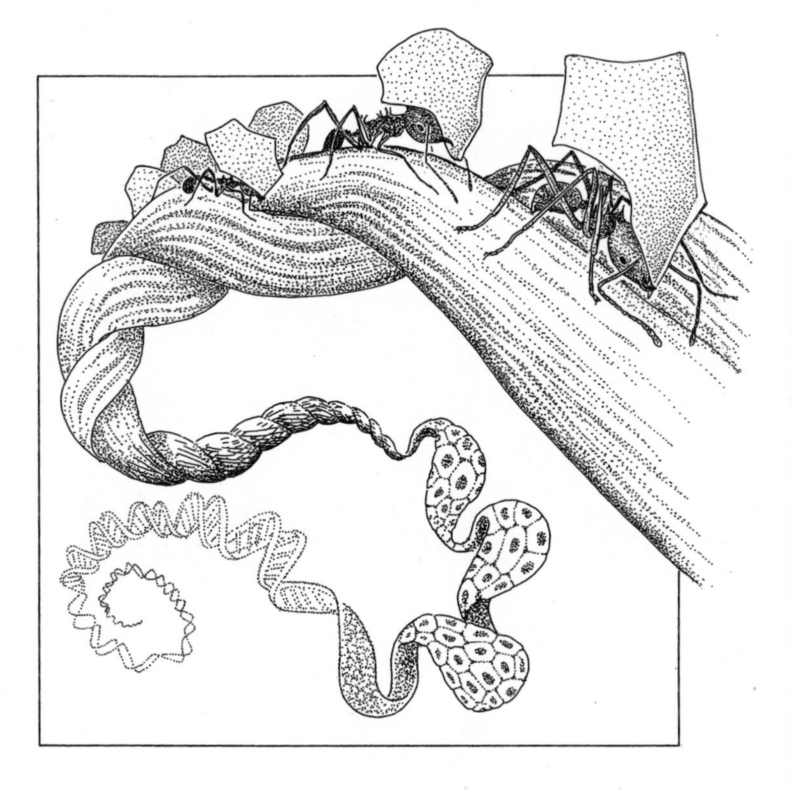

CERTAIN METAPHYSICAL constructs that historian of science Gerald Holton has called the "themata" of science are more powerful—and less vulnerable—than ordinary theories. Isaac Newton's idea of a book of Nature authored by God, Charles Darwin's vision of the grandeur of natural selection, and Friedrich Engels' description of dialectical synthesis are perhaps the most familiar examples. These metaphysical themes have shaped the direction of theory while influencing how scientists think about the totality of their life's work. It is my impression that such a thematic shift has begun to occur in biology.

The shift, if I have interpreted it correctly, will eventually move biology toward an earlier, more robust view of its raison d'être. Until the 1950s, biologists emphasized taxonomic groups of organisms, such as insects, fungi, and flowering plants, rather than levels of organization as in molecular, cell, organismal, and ecosystem biology. During the 1950s an enormously beneficial shift occurred to accommodate the beginnings of molecular and cell biology. It reflected the belief that biological laws or principles must be sought by intensive level-of-organization analysis, not in specialized study of particular kinds of organisms. That worldview, however, is now giving way to another, more balanced perception of the science of life, as follows. In the near future, while some biologists continue to think solely in terms of levels of organization and search for the broadest possible generalizations, a greater number will again commit themselves to the study of particular groups of organisms across all levels of organization. Impelling this shift is the notion that each group of organisms is of fundamental and unalterable value complete unto itself. As a result, it seems likely that the principal orientation will change from the levels of biological organization to taxonomic groups of organisms studied across all levels of organization. This shift can be visualized as a tilt from a nearly horizontal to a more vertical orientation—rotating not the full 90 degrees but, say, 45 degrees.

The result will be a pluralization of biology and the restoration of the expert naturalist to a position of leadership in biological research. By pluralization I mean increased esteem for and growth of studies of particular groups of organisms for their own sake. Put another way, taxon-oriented disciplines such as herpetology and nematology will regain ground lost to level-oriented disciplines such as cell biology and ecology. The word "fundamental," applied so freely to molecular and cellular biology, will be applied not just to broad generalizations about one or two levels of organization but also to important discoveries about individual taxa, even if the information cannot be readily applied to other taxa.

This shift is not retrograde; it won't return biology to old-fashioned, purely descriptive natural history. The technical reach of the new naturalist extends from the molecular to the populational levels, as evolutionists learn molecular techniques and molecular biologists interest themselves in the evolution of the organisms they study. As biologists increasingly commit themselves to particular groups of organisms, they seem destined to converge toward a common language and methodology. Herpetologists, nematologists, and the molecular biologists who collaborate with them have already begun to speak effectively to one another in new, shared tongues.

Why Pluralization Is Likely

The first trend suggesting such an intellectual reorganization is the growing recognition that few if any universal principles exist in biology that are both precise and widely applicable. The vast majority of investigations in molecular, cellular, and other level-oriented disciplines typically reveal truths that though jointly rooted in the physical sciences, concern only particular species or at most limited groups of species. Consider three textbook examples of basic discoveries: endocytosis by neutrophils, the action of juvenile hormones in holometabolous insects, and the density-dependent control of rodent populations. None of these findings applies beyond the taxonomic group in which it was discovered. The chief value is heuristic: the discoveries stimulate searches for parallel or analogous phenomena in broader groups of organisms. They are cited as phenomena to look for elsewhere, representatives of categories that might be abstracted into broader spheres of generality.

New general principles, the grail of biology, are becoming ever more elusive. Density dependence, for example, turns out to be present in some species and absent in others, and when present to take forms that are understandable only with a knowledge of the life cycle of each particular species and the ecosystem in which it lives. The same is true of immunochemistry, chemoreception, kin selection, and so on and on. It seems to me a remarkable feature of biology that while factual knowledge is growing exponentially, with a doubling time of perhaps 10 to 20 years, the number of broadly applicable discoveries made per investigator per year is declining steeply. A major reason for this trend is the historicity of biological phenomena, which generates special cases and crumbles generality in direct proportion to the depth of understanding.

The swift advance in knowledge achieved during the levels-directed revolution also carries within it the seeds of the decline of research confined to one or two adjacent layers of organization. Soon after new methods are invented, they are modularized—transformed into streamlined, partially automated packages and made available to all. Once-arcane techniques such as electron microscopy, amino acid sequencing, and multivariate analysis have been captured in easily followed operator's instructions for commercially available instruments. Symbiosis among the disciplines is the natural result. Systematists now routinely compare proteins, and molecular biologists construct phylogenetic trees.

At the same time, biologists are placing new emphasis on the uniqueness of each species, as something more than just a collection of related organisms and a great deal more than an interchangeable unit at the population level of organization. When you have seen one species of chrysomelid beetle, you have emphatically not seen them all. In fact you still know very little about the family Chrysomelidae. In its genes each species contains on the order of a million to a billion bits of information, assembled by an almost inconceivable number of events in mutation, recombination, and natural selection during an average life span, according to taxonomic group, of one to 10 million years. And the better each species is understood, the greater is the esteem for research conducted on it.

What does this particularity mean for the understanding of life as a whole? No one knows the number of species of living organisms, including animals, plants, and microorganisms, but there are probably at least 5 million, and the number could be as high as 100 million.

Whatever the number, it is believed to represent less than 1 percent of all the species that ever existed in geological time. We have only begun even a superficial exploration of life on Earth, living and extinct.

To the extent that this panorama is encompassed by biology, pure history will become more important. Because most biological phenomena occur in only a small portion of phyletic lines, their pattern of origin assumes an independent importance. Phylogeny (the branching patterns) and evolutionary grades (the levels of adaptation attained) form the core of biology as surely as those occasional organizational rules that hitherto have so tenuously bound biology into a unified discipline.

It also seems to follow that the more fully diversity is explored, the more quickly will the true unifying principles be discovered. The laws of biology are written in the language of diversity. Researchers often speak half-humorously about the rule attributed to the Danish physiologist August Krogh: for every biological problem there exists an organism ideal for its solution. What can be called the inverse Krogh rule applies with equal force: for every organism there exists a problem for the solution of which it is ideally suited and other problems for which it is useless. Colon bacteria are wonderful for genetic mapping but not for meiosis. Langurs and lions have given us the key to understanding infanticide but would have been a wretched first choice for genetic mapping. Every kind of organism has a place under the epistemological sun.

In short, the future of basic biological research lies

substantially in the exploration of diversity. The surest path to discovery will be systematics of a new kind, in which deep knowledge of particular groups of organisms is promoted by research that moves back and forth across all the levels of biological organization. To be a world authority on nematodes or diatoms or palms will take on a new status and entail new expertise—and also new responsibilities.

The Case of Neurobiology

Neurobiology and behavior illustrate the direction in which most of biological research is moving. The most productive strategy has proved to be the selection and detailed analysis of paradigmatic species to illuminate two or more adjacent levels of organization in the tortuous route from genes to behavior. During the past thirty years a number of such key species have risen to prominence. They range from the simplest to the most complex (including humanity), with each species being used to study some phenomenon for which it provides relatively easy or even unique access. Neurobiologists have thus demonstrated the utility of both Krogh's rule and its inverse.

At the most elementary level is the demonstration of the control of locomotion in the bacterium *Escherichia coli*. The individual bacterium moves by rotating its flagellum like the propeller of a ship. It alters course by changing the direction of the spin of the flagellum, which causes it to tumble and points it in a random new direction. By continuous trial and error it is able to move toward nutrients and away from toxic substances. Partly because of the simplicity of the system, biologists have made remarkable progress in identifying the proteins that recognize chemical stimuli, as well as the central processing proteins. The genes prescribing the key proteins have been located. The extreme simplicity of the behavioral system has thus permitted biologists to characterize behavior all the way down to the level of the genes, but the behavioral patterns can be analogized with only a minute part of any pattern in more complex organisms.

Biologists are also making rapid progress in the genetic dissection of more complicated organisms, the *Drosophila* fruit flies, particularly *Drosophila melanogaster*, because these insects are relatively easy to manipulate genetically. Researchers can even create individuals that are mosaics of male and female cells. These gynandromorphs are used to locate the sensory and nervous tissues that mediate certain forms of reproductive behavior. Investigators have been able to correlate the sex of the various tissues and the behaviors of the individual fly and thence locate genes for the processing of sensory information and efferent commands through the nervous system. Other research has led to the identification of many genes that control mating and orientation, as well as a portion of the molecular pathways that lead to the behavioral phenotype.

Sophisticated neurophysiological recordings have also traced in considerable detail the roles and patterns of discharge of individual neurons in the sea snail *Aplysia californica*. The cellular basis and, increasingly, the molecular mechanisms of elementary forms of learning have begun to yield to this approach, which takes advantage of the relative accessibility and anatomical simplicity of the sea snail's nervous system.

At a still higher level of organization, the social insects provide instructive paradigms. In the ants, bees, wasps, and termites, most behavior has no meaning except when fitted into a total pattern of different responses on the part of other colony members. One of the most dramatic examples are African driver ants of the genus *Dorylus*. The colony consists of a single queen and as many as twenty million workers. This hexapod empire is held together by powerful attractants and ovary-suppressing pheromones produced by the queen. The workers use numerous procedures in chemical communication to recruit nestmates for different functions. Various combinations of chemical and tactile signals are employed to direct nestmates to food discoveries, new territory, and new nest sites. Although each ant performs fewer than fifty behavioral acts, the caste system and division of labor permit a complicated and effective repertoire at the level of the colony. Driver ant colonies and other insect societies can be legitimately viewed as superorganisms. They can be taken apart, analyzed, and reassembled in much the same manner as the bacterium and *Drosophila*, and with considerably greater ease, to illustrate what may eventually prove to be some of the more general features of biological organization.

Neurobiology and behavioral biology should continue to progress through the skillful use of the comparative method across all levels of organization. Again, the dominant theme in the new approach is pluralism. Particular species are chosen for the levels of organization to which they provide easiest access. The whole data base from laboratory and field observations can be fitted together only when the idiosyncrasies of each species are interpreted in terms of its evolutionary history and placed in the ecosystem. Enduring biological principles will emerge when enough of this information has accumulated. How much information will in fact be needed before we can speak of a complete theory of biology can only be guessed.

The Stewardship of Systematics

Deep expertise in particular groups of organisms, if combined with freewheeling opportunism in the choice of problems, is the wave of the future for biology. Individual investigators will range more and more easily from molecule to population, piecing together the clearest images taken from different species of organisms to create a synthesis that will, like an increasingly detailed mosaic, constitute modern biology. It seems equally likely that the progress of the science must depend on the spread of expertise across more and more species, with a full picture of the world biota as the ultimate goal. To intensify pluralism in this fashion requires the renaissance of systematics as a principal orienting mode of biology.

The trend toward pluralism places a special obligation on systematists, including the core subset of researchers called taxonomists. As an expert on a group of species, a *systematist* is interested primarily in diversity, including classification, but ranges freely into other aspects of the biology of the favored group. A *taxonomist* is a systematist who is responsible for so many species that he has time only for their classification.

The resources of museums and other institutions housing major collections are already strained in the service of pure taxonomy, quite apart from the wider adventure of systematics. Aside from a few highly exceptional taxa such as mammals and birds, the identification of newly discovered organisms is often delayed for months or years. For many groups of organisms there are no experts at all. Taxonomists, and hence the broader class of systematists, cannot do the job expected of them. As biodiversity studies develop and practical needs also proliferate, especially in the tropics, this shortfall could become critical within a very few years.

Systematists themselves are currently in the thrall of

what some critics perceive as largely sterile arguments over the methodology of phylogenetic reconstruction. I see this methodological phase as invigorating and productive, even though it is certain to be greatly augmented or in some cases even replaced by direct reading of the genetic codes. The debates have produced sound techniques for inferring degrees of similarity and branching during species formation. Even more important, they have resulted in a remarkable improvement in taxonomic procedures, amounting to the standardization of techniques by which results can be replicated and independently tested. But most of this activity is after all just methodology. The time has come for systematics to move on, to meet its destiny. Otherwise, why was all the work done?

For that matter, why does a biologist do research at all? To discover, of course. Alfred North Whitehead said that a scientist does not discover in order to know; he knows in order to discover. But in biology there is much more to the discovery impulse. The uniqueness of phylogenetic lineages makes history all-important, and history in turn generates a sense of the sacredness of place and of life—not general and abstract but concerned with a single organism in a particular habitat, watched for an explicit span of time. Thus biology satisfies the two great expansive drives of the human mind: exploration and intellectual enrichment. The governing concept of pluralism ensures that biology will never, within any time imaginable, exhaust these two drives.

Systematists who are experts on particular groups of

organisms, as opposed to those who study method alone, must overcome a certain reticence in their relationship to the rest of science. Too often I have heard other biologists say that a taxonomist with a grant goes off somewhere and writes a specialized monograph, and that is the end of it. And they say that systematists have not formulated a set of central questions in science that they alone are peculiarly qualified to answer.

If systematics were really an exhausted relic from the premolecular age, we should not try to keep it from entering the long sleep of senescence. But the exact opposite is the case. As it was during a glorious past and will be again, systematics in the broad sense is the key to the future of biology.

The responsible expert is the steward of a chosen taxonomic group in the service of science. He or she knows best which organisms exist and where, which are most endangered, which offer new kinds of problems to be solved, and which are most likely to benefit humankind. The systematist's best strategy is to explain such matters to as broad an audience as possible while inviting the collaboration of other biologists. No one but the systematist can reveal the particular and extraordinary value of the alcyonacean corals, chytrid fungi, anthribid weevils, sclerogibbid wasps, melostomes, ricinuleids, elephant fish, and so on down the long and enchanted roster.

BIOPHILIA AND THE ENVIRONMENTAL ETHIC

BIOPHILIA, IF IT EXISTS, and I believe it exists, is the innately emotional affiliation of human beings to other living organisms. From the scant evidence concerning its nature, biophilia is not a single instinct but a complex of learning rules that can be teased apart and analyzed individually. The feelings molded by the learning rules fall along several emotional spectra, from attraction to aversion, awe to indifference, and peacefulness to fear-driven anxiety. These multiple strands of emotional response are woven into symbols composing a large part of culture. When human beings remove themselves from the natural environment, the biophilic learning rules are not replaced by modern versions equally well adapted to contemporary technological features of life. Instead, they persist from generation to generation, atrophied and fitfully manifested in the artificial new environments. It is no accident of culture that more children and adults visit zoos than attend all major professional sports combined (at least in the United States and Canada), that the wealthy continue to seek dwellings on prominences above water amidst parkland,

and that urban dwellers continue to dream of snakes for reasons they cannot explain.

Were there no evidence of biophilia at all, the hypothesis of its existence would still be compelled by pure evolutionary logic. The reason is that human history did not begin a mere 8,000 or 10,000 years ago, with the invention of agriculture and villages. It began hundreds of thousands or millions of years ago, with the origin of the genus *Homo*. For more than 99 percent of human history people have lived in hunter-gatherer bands intimately involved with other organisms. During this period of deep history, and still farther back, into paleohominid times, they depended on an exact learned knowledge of crucial aspects of natural history. That much is true even of chimpanzees today, who use primitive tools and have a practical knowledge of plants and animals. As language and culture expanded, humans also used living organisms of diverse kinds as a principal source of metaphor and myth. In short, the brain evolved in a biocentric world, not a machine-regulated one. It would therefore be quite extraordinary to find that all learning rules related to that world had been erased in a few thousand years, even in the tiny minority of humans who have existed for more than one or two generations in wholly urban environments.

The significance of biophilia in human biology is potentially profound, even if it exists solely as weak learning rules. It is relevant to our thinking about nature, about the landscape, the arts, and mythopoeia, and it invites us to take a new look at environmental ethics.

How could biophilia have evolved? The likely answer is biocultural evolution, during which culture was elaborated under the influence of hereditary learning propensities while the genes prescribing the propensities were spread by natural selection in a cultural context. The learning rules can be inaugurated and fine-tuned variously by an adjustment of sensory thresholds, by a quickening or blockage of learning, and by modification of emotional responses. Charles Lumsden and I have envisioned biocultural evolution to be of a particular kind, gene-culture coevolution, which traces a spiral trajectory through time: a certain genotype makes a behavioral response more likely, the response enhances survival and reproductive fitness, the genotype consequently spreads through the population, and the behavioral response grows more frequent. Add to this the strong general tendency of human beings to translate emotions into myriad dreams and narratives, and the necessary conditions are in place to cut the historical channels of art and religious belief.

Gene-culture coevolution is a plausible explanation for the origin of biophilia. The hypothesis can be made explicit by the human relation to snakes. The sequence I envision is the following, drawn principally from elements established by the art historian and biologist Balaji Mundkur:

- Poisonous snakes cause sickness and death in primates and other mammals throughout the world.
- Old World monkeys and apes generally combine a strong natural fear of snakes with fascination for these

animals and the use of vocal communication, the latter including specialized sounds in a few species, all drawing attention of the group to the presence of snakes in the near vicinity. Thus alerted, the group follows the intruders until they leave.

- Human beings are also genetically averse to snakes. They are quick to develop fear and even full-blown phobias with very little negative reinforcement. (Other phobic elements in the natural environment include dogs, spiders, closed spaces, running water, and heights. Few if any modern artifacts are as effective, including even those most dangerous, such as guns, knives, automobiles, and electric wires.)
- In a manner true to their status as Old World primates, human beings are also fascinated by snakes. They pay admission to see captive specimens in zoos. They employ snakes profusely as metaphors and weave them into stories, myth, and religious symbolism. The serpent gods of cultures they have conceived all around the world are furthermore typically ambivalent. Often semihuman in form, they are poised to inflict vengeful death but also to bestow knowledge and power.
- People in diverse cultures dream more about serpents than about any other kind of animal, conjuring as they do so a rich medley of dread and magical power. When shamans and religious prophets report such images, they invest them with mystery and symbolic authority. In what seems to be a logical consequence, serpents are

also prominent agents in mythology and religion in a majority of cultures.

Here then is the ophidian version of biophilia hypothesis expressed in briefest form: constant exposure through evolutionary time to the malign influence of snakes, the repeated experience encoded by natural selection as a hereditary aversion and fascination, which in turn is manifested in the dreams and stories of evolving cultures. I would expect that other biophilic responses have originated more or less independently, by the same means but under different selection pressures and with the involvement of different gene ensembles and brain circuitry.

Of course, this formulation is fair enough as a working hypothesis, but we must also ask how such elements can be distinguished, and how the general biophilia hypothesis might be tested. One mode of analysis, reported by Jared Diamond, is the correlative analysis of knowledge and attitude of peoples in diverse cultures, designed to search for common denominators in the total human pattern of response. Another, advanced by Robert Ulrich and other psychologists, is the precisely replicated measurement of the physiological responses of human subjects to both attractive and aversive natural phenomena. This direct psychological approach can be made increasingly persuasive, whether for or against a biological bias, when two elements are added. The first is the measurement of heritability in the intensity of the responses to the psychological tests used. The second element is the tracing of cognitive development in children to identify key stimuli that evoke the responses, along with the ages of maximum sensitivity and learning propensity. For example, the slithering motion of an elongate form appears to be the key stimulus producing snake aversion, and preadolescence may be the maximally sensitive period for acquiring the aversion.

Given that humanity's relation to the natural environment is as much a part of deep history as social behavior itself, cognitive psychologists have been strangely slow to address its mental consequences. Our ignorance could be regarded as just one more blank space on the map of academic science, awaiting genius and initiative, except for one important circumstance: the natural environment is disappearing. As a consequence, psychologists and other scholars are obligated to consider biophilia in more urgent terms. What, they should ask, will happen to the human psyche when such a defining part of the human evolutionary experience is diminished or erased?

There is no question in my mind that the most harmful part of ongoing environmental despoliation is the loss of biodiversity. The reason is that the variety of organisms, from alleles (differing gene forms) to species, once lost, cannot be regained. If diversity is sustained in wild ecosystems, the biosphere can be recovered and used by future generations to any degree desired and with benefits literally beyond measure. To the extent it is diminished, humanity will be poorer for all generations to come. How much poorer? The following estimates give a rough idea:

- Consider first the question of the amount of biodiversity. The number of species of organisms on Earth is unknown to the nearest order of magnitude. About 1.5 million species have been given names to date, but the actual number is likely to lie somewhere between 10 and 100 million. Among the least-known groups are the fungi, with 69,000 known species but 1.6 million thought to exist. Also poorly explored are at least several million and possibly tens of millions of species of arthropods in the tropical rain forests; and millions of invertebrate species on the vast floor of the deep sea. The true black hole of systematics, however, may be bacteria. Although roughly 4,000 species have been formally described, recent studies in Norway have indicated the presence of from 4,000 to 5,000 species, almost all new to science, among the 10 billion individual organisms found on average in each gram of forest soil, and another 4,000 to 5,000 species, different from the first set and also mostly new, in an average gram of nearby marine sediments.
- Fossil records of marine invertebrates, African ungulates, and flowering plants indicate that under natural conditions each clade—a species and its descendants—lasts an average of 500,000 to 10 million years. The longevity is measured from the time the ancestral form splits off from its sister species to the time of the extinction of the last descendant. It varies according to the group of organisms. Mammals, for example, are shorter-lived than invertebrates.

- Bacteria contain on the order of a million nucleotide pairs in their genetic code, and more complex (eukaryotic) organisms from algae to flowering plants and mammals contain one to 10 billion nucleotide pairs.
- Because of their great age and genetic complexity, species are exquisitely adapted to the ecosystems in which they live.
- The number of species on Earth is being reduced by a rate from 100 to 1,000 times higher than in prehuman times. The current removal rate of tropical rain forest, over 1 percent of cover each year, translates (if we use the most conservative parameter value) to approximately 0.3 percent of the species extirpated immediately or at least doomed to much earlier extinction than would otherwise have been the case. Most systematists with global expressions believe that more than half the species of organisms on earth live in the tropical rain forests. If there are 10 million species in these habitats—a conservative estimate—the rate of loss may be 30,000 a year, 74 a day, 3 an hour. This rate, though horrendous, is actually a minimal estimate, in the sense that it is based on the area-species relation alone. It does not take into account extinction due to pollution, disturbance short of clear-cutting, and the introduction of exotic species.

Other species-rich habitats, including coral reefs, river systems, lakes, and Mediterranean-type heathland, are under similar assault. When the final remnants of such habitats are destroyed in a region—the last of the ridges on a mountainside cleared, for example, or the last riffles flooded by a downstream dam—species are wiped out en masse. The first 90 percent reduction in area of a habitat lowers the species number by one-half. The final 10 percent eliminates the second half.

It is a guess, subjective but very defensible, that if the current rate of habitat alteration continues unchecked, 20 percent or more of the Earth's species will disappear or be consigned to early extinction through human action taken during the next thirty years. From prehistory to the present time humanity has probably already eliminated 10 or even 20 percent of the species. The number of bird species, for example, is down by an estimated 25 percent, from 12,000 to 9,000, with a disproportionate share of the losses occurring on islands. Most of the megafaunas—the largest mammals and birds—appear to have been destroyed in more remote parts of the world by the first wave of hunter-gatherers and agriculturists millennia ago. The loss of plants and invertebrates is likely to have been much smaller, but studies of archaeological and other subfossil deposits are too few to permit even a crude estimate. The human impact, from prehistory to the present time and projected into the next several decades, threatens to be the greatest extinction spasm since the end of the Mesozoic era, 65 million years ago.

Assume, for the sake of argument, that 10 percent of the world's species that existed just before the advent of humanity are already gone, and that another 20 percent are destined to vanish quickly unless drastic action is taken. The fraction lost—and it will be a great deal no matter what action is taken—cannot be replaced by evolution in any period that has meaning for the human mind. Following each of the five previous major spasms of the past 550 million years, life required about 10 million years of natural evolution to recover. What humanity is doing now in a single lifetime will impoverish our descendants for virtually all time to come. Yet critics often respond, "So what? If only half the species survive, that is still a lot of biodiversity—is it not?"

The answer most frequently urged right now by conservationists, myself among them, is that the vast material wealth offered by biodiversity is at risk. Wild species are an untapped source of new pharmaceuticals, crops, fibers, pulp, petroleum substitutes, and agents for the restoration of soil and water. This argument is demonstrably true—and it certainly tends to stop anti-conservation libertarians in their tracks—but it contains a dangerous practical flaw when relied upon exclusively. If species are to be judged by their potential material value, they can be priced, traded off against other sources of wealth, and—when the price is right—discarded. Yet who can judge the *ultimate* value of any particular species to humanity? Whether the species offers immediate advantage or not, no means exist to measure what benefits it will offer during future centuries of study, what scientific knowledge, or what service to the human spirit.

At last I have come to the word so hard to express,

spirit. With reference to the spirit we arrive at the connection between biophilia and the environmental ethic. The great philosophical divide in moral reasoning about the remainder of life is whether or not other species have an innate right to exist. That decision rests in turn on the most fundamental question of all, whether moral values exist apart from humanity, in the same manner as mathematical laws, or whether they are idiosyncratic constructs that evolved in the human mind through natural selection, and thus of the spirit. Had a species other than humans attained high intelligence and culture, it would probably have fashioned different moral values. Civilized termites, for example, would support cannibalism of the sick and injured, eschew personal reproduction, and make a sacrament of the exchange and consumption of feces. The termite "spirit," in short, would have been immensely different from the human spirit, horrifying to us in fact. The constructs of moral reasoning, in this evolutionary view, are the learning rules, the propensities to acquire or to resist certain emotions and kinds of knowledge. They have evolved genetically because they confer survival and reproduction on human beings.

The first of the two alternative propositions—that species have universal and independent rights, regardless of how else human beings feel about the matter—may be true. To the extent the proposition is accepted, it will certainly steel the determination of environmentalists to preserve the remainder of life. But the species-right argument alone, like the materialistic argument alone, is a dangerous gambit on which to risk biodiversity. Its reasoning, for all its directness and power, remains intuitive, aprioristic, and lacking in objective evidence. Who but humanity, it can be immediately asked, gives such rights? Where is the enabling canon written? And such rights, even if granted, are always subject to rank-ordering and relaxation. A simplistic adjuration for the right of a species to live can be answered by a simplistic call for the right of people to live. If a last section of forest needs to be cut to perpetuate the survival of a local economy, the rights of the myriad species in the forest may be cheerfully recognized but given a lower and fatal priority.

Without attempting to resolve the issue of the innate rights of species, I will argue the necessity of a robust and richly textured anthropocentric ethic apart from the issue of rights, one based on the hereditary needs of our own species. In addition to the well-documented utilitarian potential of wild species, the diversity of life has immense aesthetic and spiritual value. The ideas outlined below are already familiar to many conservationists and ethicists, yet the evolutionary logic is still relatively new and poorly explored, and therein lies the challenge to scientists and other scholars.

Biodiversity is the Creation. Ten million or more species are still alive, each defined by up to billions of nucleotide pairs and a far larger, in fact astronomical, number of possible genetic recombinants. These constitute the arena in which evolution continues to occur. Despite the fact that living organisms compose a mere one ten-billionth part of the mass of Earth, biodiversity is the most information-rich part of the known universe. More organization and complexity exist in a handful of soil than on the surfaces of all the other planets combined. If humanity is to have a satisfying creation myth consistent with scientific knowledge—a myth that itself seems to be an essential part of the human spirit—the narrative will find its starting point in the origin of the diversity of life.

Other species are our kin. This perception is literally true in evolutionary time. All higher eukaryotic organisms, from flowering plants to insects and humanity itself, are thought to have descended from a single ancestral population that lived about 1.8 billion years ago. Single-celled eukaryotes and bacteria are linked by still more remote ancestors. This distant kinship is stamped by a common genetic code and elementary features of cell structure. Humanity did not soft-land into the teeming biosphere like an alien from another planet. We arose from other organisms already here, whose great diversity, conducting experiment upon experiment in the production of new life forms, eventually hit upon the human species.

The biodiversity of a country is part of its national heritage. Each country in turn possesses its own unique assemblages of plants and animals, including, in almost all cases, species and geographic races found nowhere else. Those assemblages are the product of the deep history of the national territory, extending back long before the coming of man.

Biodiversity is the frontier of the future. Humanity needs a vision of an expanding and unending future. That spiritual craving cannot be satisfied by the colonization of space. The other planets are inhospitable and immensely expensive to reach. The nearest stars are so far away that voyagers would need thousands of years just to report back. The true frontier for humanity is life on Earth—its exploration and the transport of knowledge about it into science, art, and practical affairs. The circumstances that validate the proposition are, to repeat briefly: 90 percent or more of the species of plants, animals, and microorganisms lack even so much as a scientific name; each of the species is immensely old by human standards and has been wonderfully molded to its environment; life around us exceeds in complexity and beauty anything else humanity is ever likely to encounter.

The manifold ways by which human beings are tied to the remainder of life are very poorly understood, crying for new scientific inquiry and bold aesthetic interpretation. The portmanteau terms "biophilia" and "biophilia hypothesis" will serve well if they do no more than call attention to psychological phenomena that rose from deep human history, stemmed from interaction with the natural environment, and are now quite likely resident in the genes themselves. The search is rendered more urgent by the rapid disappearance of the living part of that environment, creating a need not only for a better understanding of human nature but for a more powerful and intellectually convincing environmental ethic based upon it.

IS HUMANITY SUICIDAL?

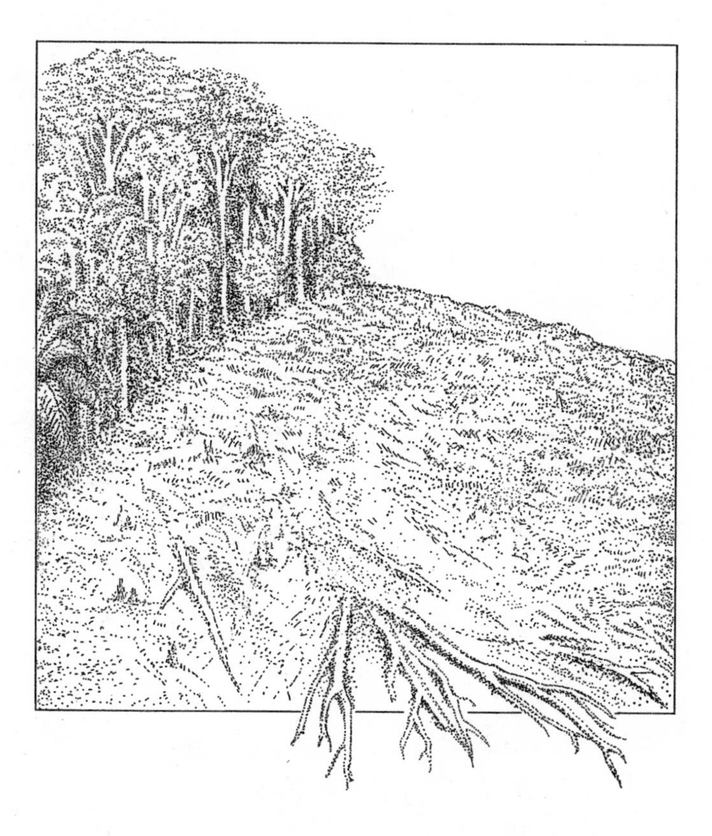

IMAGINE THAT ON AN ICY moon of Jupiter—say, Ganymede—the space station of an alien civilization is concealed. For millions of years its scientists have closely watched Earth. Because their law prevents settlement on a living planet, they have tracked the surface by means of satellites equipped with sophisticated sensors, mapping the spread of large assemblages of organisms, from forests, grasslands, and tundras to coral reefs and the vast planktonic meadows of the sea. They have recorded millennial cycles in the climate, interrupted by the advance and retreat of glaciers and scattershot volcanic eruptions.

The watchers have been waiting for what might be called the Moment. When it comes, occupying only a few centuries and thus a mere tick in geological time, the forests shrink back to less than half their original cover. Atmospheric carbon dioxide rises to the highest level in 100,000 years. The ozone layer of the stratosphere thins, and holes open at the poles. Plumes of nitrous oxide and other toxins rise from fires in South America and Africa, settle in the upper troposphere, and drift eastward across the oceans. At night the land surface brightens with millions of pinpoints of light, which coalesce into blazing swaths across Europe, Japan, and eastern North America. A semicircle of fire spreads from gas flares around the Persian Gulf.

It was all but inevitable, the watchers might tell us if we met them, that from the great diversity of large animals, one species or another would eventually gain intelligent control of Earth. That role has fallen to *Homo sapiens*, a primate risen in Africa from a lineage that split away from the chimpanzee line 5 to 8 million years ago. Unlike any creature that lived before, we have become a geophysical force, swiftly changing the atmosphere and climate as well as the composition of the world's fauna and flora. Now in the midst of a population explosion, the human species has doubled to 5.5 billion during the past 50 years. It is scheduled to double again in the next 50 years. No other single species in evolutionary history has even remotely approached the sheer mass in protoplasm generated by humanity.

Darwin's dice have rolled badly for Earth. It was a misfortune for the living world in particular, many scientists believe, that a carnivorous primate and not some more benign form of animal made the breakthrough. Our species retains hereditary traits that add greatly to our destructive impact. We are tribal and aggressively territorial, intent on private space beyond minimal requirements, and oriented by selfish sexual and reproductive drives. Cooperation beyond the family and tribal levels comes hard.

Worse, our liking for meat causes us to use the sun's energy at low efficiency. It is a general rule of ecology that (very roughly) only about 10 percent of the sun's energy captured by photosynthesis to produce plant tissue is converted into energy in the tissue of herbivores, the animals that eat the plants. Of that amount, 10 percent reaches the tissue of the carnivores feeding on the herbivores. Similarly, only 10 percent is transferred to carnivores. And so on for another step or two. In a wetlands chain that runs from marsh grass to grasshopper to warbler to hawk, the energy captured during green production shrinks a thousandfold.

In other words, it takes a great deal of grass to support a hawk. Human beings, like hawks, are top carnivores, at the end of the food chain whenever they eat meat, two or more links removed from the plants; if chicken, for example, two links; and if tuna, four links. Even with most societies confined today to a largely vegetarian diet, humanity is gobbling up a large part of the rest of the living world. We appropriate between 20 and 40 percent of the sun's energy that would otherwise be fixed into the tissue of natural vegetation, principally by our consumption of crops and timber, construction of buildings and roadways, and creation of wastelands. In the relentless search for more food we have reduced animal life in lakes, rivers, and now, increasingly, the open ocean. And everywhere we pollute the air and water, lower water tables, and extinguish species.

The human species is, in a word, an environmental hazard. It is possible that intelligence in the wrong kind of species was foreordained to be a fatal combination for the biosphere. Perhaps a law of evolution is that intelligence usually extinguishes itself.

This admittedly dour scenario is based on what can be termed the juggernaut theory of human nature, which holds that people are programmed by their genetic heritage to be so selfish that a sense of global responsibility will come too late. Individuals place themselves first, family second, tribe third, and the rest of the world a distant fourth. Their genes also predispose them to plan ahead for at most one or two generations. They fret over the petty problems and conflicts of their daily lives and respond swiftly and often ferociously to slight challenges to their status and tribal security. But oddly, as psychologists have discovered, humans also tend to underestimate both the likelihood and impact of such natural disasters as major earthquakes and great storms.

The reason for this myopic fog, evolutionary biologists contend, is that it was actually advantageous during all but the last few millennia of the 2 million years of existence of the genus *Homo*. The brain evolved into its present form during this long stretch of evolutionary time, during which humans existed in small, preliterate hunter-gatherer bands. Life was precarious and short. A premium was placed on close attention to the near future and early reproduction, and little else. Disasters of a magnitude that occurred only once every few centuries were forgotten or transmuted into myth. So today the human mind still works comfortably backward and forward for only a few years, spanning a period not exceeding one or two generations. Those in past ages whose genes inclined them to short-term thinking lived longer and had more children than those who did not. Prophets never enjoyed a Darwinian edge.

The rules have recently changed, however. Global crises are proliferating within the life span of the generation now coming of age, a foreshortening that may explain why young people express more concern about the environment than do their elders. The time scale has contracted because of the exponential growth in both the human population and technologies impacting the environment. Exponential growth is basically the same as the increase of wealth by compound interest. The larger the population, the faster the growth; the faster the growth, the sooner the population becomes still larger. In Nigeria, to cite one of our more fecund nations, the population is expected to double from its 1988 level to 216 million by the year 2010. If the same rate of growth were to continue to 2110, Nigeria's population would exceed that of the entire present population of the world.

With people everywhere seeking a better quality of life, the search for resources is expanding even faster than the population. The demand is being met by an increase in scientific knowledge, which doubles every 10 to 15 years. It is accelerated further by a parallel rise in environment-devouring technology. Because Earth is finite in many resources that determine the quality of life—including arable soil, nutrients, fresh water, and space for natural ecosystems—doubling of consumption at constant intervals can bring disaster with shocking suddenness. Even when a nonrenewable resource has been only half used, it is still only one interval away from the end. Ecologists like to make this point with the French riddle of the lily pond. At first there is only one lily pad in the pond, but the next day it doubles, and thereafter each of its descendants doubles. The pond completely fills with lily pads in 30 days. When is the pond exactly half full? Answer: on the twenty-ninth day.

Yet, mathematical exercises aside, who can safely measure the human capacity to overcome the perceived limits of Earth? The question of central interest is this: Are we racing to the brink of an abyss, or are we just gathering speed for a takeoff to a wonderful future? The crystal ball is clouded; the human condition baffles all the more because it is both unprecedented and bizarre, almost beyond understanding.

In the midst of uncertainty, opinions on the human prospect have tended to fall loosely into two schools. The first, exemptionalism, holds that since humankind is transcendent in intelligence and spirit, our species must have been released from the iron laws of ecology that bind all other species. However serious the problem, civilized human beings, by ingenuity, force of will, and—who knows—divine dispensation, will find a solution.

Population growth? Good for the economy, claim some of the exemptionalists, and in any case a basic human right, so let it run. Land shortages? Try fusion energy to power the desalting of seawater, then reclaim the world's deserts. (The process might be assisted by towing icebergs to coastal pipelines.) Species going extinct? Not to worry. That is nature's way. Think of humankind as only the latest in a long line of exterminating agents in geological time. In any case, because our species has pulled free of old-style, mindless Nature, we have begun a different order of life. Evolution should now be allowed to proceed along this new trajectory. Finally, resources? The planet has more than enough resources to last indefinitely, if human genius is allowed to address each new problem in turn, without alarmist and unreasonable restrictions imposed on economic development. So hold the course, and touch the brakes lightly.

The opposing idea of reality is environmentalism, which sees humanity as a biological species tightly dependent on the natural world. However formidable our intellect may be and however fierce our spirit, the argument goes, those qualities are not enough to free us from the constraints of the natural environment in which the human ancestors evolved. We cannot draw confidence from successful solutions to the smaller problems of the past. Many of Earth's vital resources are about to be exhausted, its atmospheric chemistry is deteriorating, and human populations have already grown dangerously large. Natural ecosystems, the wellsprings of a healthful environment, are being irreversibly degraded.

At the heart of the environmentalist worldview is the conviction that human physical and spiritual health depends on sustaining the planet in a relatively unaltered state. Earth is our home in the full, genetic sense, where humanity and its ancestors existed for all the millions of years of their evolution. Natural ecosystems—forests, coral reefs, marine blue waters—maintain the world exactly as we would wish it to be maintained. When we debase the global environment and extinguish the variety of life, we are dismantling a support system that is too complex to understand, let alone replace, in the foreseeable future. Space scientists theorize the existence of a virtually unlimited array of other planetary environments, almost all of which are uncongenial to human life. Our own Mother Earth, lately called Gaia, is a specialized conglomerate of organisms and the physical environment they create on a day-to-day basis, which can be destabilized and turned lethal by careless activity. We run the risk, conclude the environmentalists, of beaching ourselves upon alien shores like a great confused pod of pilot whales.

If my tone here has not already made my own position

clear, I now explicitly place myself in the environmentalist school. I am not so radical as to wish for a turning back of the clock; I am not given to driving spikes into Douglas firs to prevent logging; and I am distinctly uneasy with such hybrid movements as ecofeminism, which holds that Mother Earth is a nurturing home for all life and should be revered and loved as in premodern (paleolithic and archaic) societies and that ecosystem abuse is rooted in androcentric—that is, male-dominated—concepts, values, and institutions. Still, product of androcentric culture though I am, I am radical enough to take seriously the question heard with increasing frequency: Is humanity suicidal? Is the drive to environmental conquest and self-propagation embedded so deeply in our genes as to be unstoppable?

My short answer—opinion, if you wish—is that humanity is not suicidal, at least not in the sense just stated. We are smart enough and have time enough to avoid an environmental catastrophe of civilization-threatening dimensions. But the technical problems are sufficiently formidable to require a redirection of much of science and technology, and the ethical issues are so basic as to force a reconsideration of our self-image as a species.

There are reasons for optimism, reasons to believe that we have entered what might someday be generously called the Century of the Environment. The United Nations Conference on Environment and Development, held in Rio de Janeiro in June 1992, attracted more than 120 heads of government, the largest number ever assembled, and helped move environmental issues closer to the political center stage; on November 18, 1992, more than 1,500 senior scientists from 69 countries issued a "Warning to Humanity," stating that overpopulation and environmental deterioration have put the very future of life at risk. The greening of religion has become a global trend, with theologians and religious leaders addressing environmental problems as a moral issue. In May 1992, leaders of most of the major American denominations met with scientists as guests of the U.S. Senate to formulate a "Joint Appeal by Religion and Science for the Environment." Conservation of biodiversity is increasingly seen by national governments and major landowners alike as important to their country's future. Indonesia, home to a large share of the native Asian plant and animal species, has begun to shift to land-management practices that conserve and sustainably develop the remaining rain forests. Costa Rica has created a National Institute of Biodiversity. A Pan-African institute for biodiversity research and management has been founded, with headquarters in Zimbabwe.

Finally, there are favorable demographic signs. The rate of population increase is declining on all continents, although it is still well above zero almost everywhere and remains especially high in sub-Saharan Africa. Despite entrenched traditions and religious beliefs, the desire to use contraceptives in family planning is spreading.

Demographers estimate that if the demand were fully met, contraceptive use alone would reduce the eventual stabilized population by more than 2 billion.

In summary, the will is there. Yet the awful truth remains that a large part of humanity will suffer no matter what is done. The number of people living in absolute poverty has risen during the past 20 years to nearly one billion and is expected to increase another 100 million by the end of the decade. Whatever progress has been made in the developing countries, and that includes an overall improvement in the average standard of living, is threatened by a continuance of rapid population growth and the deterioration of forests and arable soil.

Our hopes must be chastened further still, and this is the central issue, by a key and seldom-recognized distinction between the nonliving and living environments. Science and the political process can be adapted to manage the nonliving physical environment. The human hand is now upon the physical homeostat. The ozone layer can be mostly restored to the upper atmosphere by elimination of CFCs, with these substances peaking at six times the present level and then subsiding during the next half-century. Also, with procedures that will prove far more difficult and initially expensive, carbon dioxide and other greenhouse gases can be pulled back to concentrations that slow global warming.

The human hand, however, is *not* upon the biological homeostat. There is no way in sight to micromanage the natural ecosystems and the millions of species they contain. That feat might be accomplished by generations to come, but then it will be too late for the ecosystems—and perhaps for us. Despite the seemingly fathomless extent of creation, humankind has been chipping away at its diversity, and Earth is destined to become an impoverished planet within a century if present trends continue. Mass extinctions are being reported with increasing frequency in every part of the world. They include half the freshwater fishes of peninsular Malaysia, half the 41 tree snails of Oahu, 44 of the 68 shallow-water mussels of the Tennessee River shoals, as many as 90 plant species growing on the Centinela Ridge in Ecuador, and, in the United States as a whole, about 200 plant species, with another 680 species and races now classified as in danger of extinction. The main cause is the destruction of natural habitats, especially tropical forests. Close behind, especially on the Hawaiian archipelago and other islands, is the introduction of rats, pigs, beard grass, lantana, and other exotic organisms that outbreed and extirpate native species.

The few thousand biologists worldwide who specialize in diversity are aware that they can witness and report no more than a very small percentage of the extinctions actually occurring. The reason is that they have facilities to keep track of only a tiny fraction of the millions of species and a sliver of the planet's surface on a yearly basis. They have devised a general standard by which to characterize the situation: that whenever careful studies are made of habitats before and after disturbance, extinctions almost always come to light. The corollary: the great majority of extinctions are never observed. Vast numbers of species are apparently vanishing before they can be discovered and named.

There is a way, nonetheless, to estimate the rate of loss indirectly. Independent studies around the world and in fresh and marine waters have revealed a robust connection between the size of a habitat and the amount of biodiversity it contains. Even a small loss in area reduces the number of species. The relation is such that when the area of the habitat is cut to one-tenth of its original cover, the number of species eventually drops by roughly one-half. Tropical rain forests, thought to harbor a majority of Earth's species (the reason conservationists get so exercised about rain forests), are being reduced by nearly that magnitude. At the present time they occupy about the same total area as that of the forty-eight conterminous United States, representing a little less than half their original, prehistoric cover; and they are shrinking each year by well over 1 percent, an amount equal to half the state of Florida. If the typical value (that is, 90 percent area loss causes 50 percent eventual extinction) is applied, the projected loss of species due to rain forest destruction worldwide is 0.3 percent across the board for all kinds of plants, animals, and microorganisms.

When area reduction and all the other extinction agents are considered together, it is reasonable to project a reduction by 20 percent or more of the rain forest species by the year 2020, climbing to 50 percent or more by midcentury, if nothing is done to change current practice. Comparable erosion is likely in other environments now under assault, including many coral reefs and Mediterranean-type heathlands of Western Australia, South Africa, and California.

The ongoing loss will not be replaced by evolution in any period of time that has meaning for humanity. Extinction is now proceeding thousands of times faster than the production of new species. The average life span of a species and its descendants in past geological eras varied according to group (like mollusks or echinoderms or flowering plants) from about 1 million to 10 million years. During the past 500 million years, there have been five great extinction spasms comparable to the one now being inaugurated by human expansion. The latest, evidently caused by the collision with Earth of an asteroid, ended the age of reptiles 66 million years ago. In each case it took more than 10 million years of evolution to completely replenish the biodiversity lost. And that was in an otherwise undisturbed natural environment. Humanity is now destroying most of the habitats where evolution can occur.

The surviving biosphere remains the great unknown of Earth in many respects. On the practical side, it is hard even to imagine what other species have to offer in the way of new pharmaceuticals, crops, fibers, petroleum substitutes, and other products. We have only a poor grasp of the ecosystem services by which other organisms cleanse the water, turn soil into a fertile living cover, and manufacture the very air we breathe. We sense but do not fully understand what the highly diverse natural world means to our aesthetic pleasure and mental well-being.

Scientists are unprepared to manage a declining biosphere. To illustrate, consider the following mission they might be given. The last remnant of a rain forest is about to be cut over. Environmentalists are stymied. The contracts have been signed, and local landowners and politicians are intransigent. In a final desperate move, a team of biologists is scrambled in an attempt to preserve the biodiversity by extraordinary means. Their assignment is the following: collect samples of all the species of organisms quickly, before the cutting starts; maintain the species in zoos, gardens, and laboratory cultures or else deep-freeze samples of the tissues in liquid nitrogen; and finally, establish the procedure by which the entire community can be reassembled on empty ground at a later date, when social and economic conditions have improved.

The biologists cannot accomplish this task, not if thousands of them came with a billion-dollar budget. They cannot even imagine how to do it. In the forest patch live legions of species: perhaps 300 birds, 500 butterflies, 200 ants, 50,000 beetles, 1,000 trees, 5,000 fungi, tens of thousands of bacteria, and so on down a long roster of major groups. Each species occupies a precise niche, demanding a certain place, an exact microclimate, particular nutrients and temperature and humidity cycles with specified timing to trigger phases of the life cycle. Many, perhaps most, of the species are locked in symbioses with other species; they cannot survive and reproduce unless arrayed with their partners in the correct idiosyncratic configurations.

Even if the biologists pulled off the taxonomic equivalent of the Manhattan Project in reverse, sorting and preserving cultures of all the species, they could not then put the community back together again. It would be like unscrambling an egg with a pair of spoons. The biology of the microorganisms needed to reanimate the soil would be mostly unknown. The pollinators of most of the flowers and the correct timing of their appearance could only be guessed. The "assembly rules," the sequence in which species must be allowed to colonize in order to coexist indefinitely, would remain in the realm of theory.

In its neglect of the rest of life, exemptionalism fails definitively. To move ahead as though scientific and entrepreneurial genius will solve each crisis that arises implies that the declining biosphere can be similarly manipulated. But the world is too complicated to be turned into a garden. There is no biological homeostat that can be worked by humanity; to believe otherwise is to risk reducing a large part of Earth to a wasteland.

The environmentalist vision, prudential and less exuberant than exemptionalism, is closer to reality. It sees humanity entering a bottleneck unique in history, constricted by population and economic pressures. In order to pass through to the other side, within perhaps 50 to 100 years, more science and entrepreneurship will have to be devoted to stabilizing the global environment. That can be accomplished, according to expert consensus, only by halting population growth and devising a wiser use of resources than has been accomplished to date. And wise use for the living world in particular means preserving the surviving ecosystems, micromanaging them only enough to save the biodiversity they contain, until such time as they can be understood and employed in the fullest sense for human benefit.

ACKNOWLEDGMENT OF SOURCES

The essays of this volume have been reprinted, in most cases with modest alterations to bring their substance up-to-date, and with permission of the publishers, from the book chapters and journal articles listed below.

"The Serpent," from the chapter of that title in *Biophilia* (Cambridge, Mass.: Harvard University Press, 1984), pp. 83–101. Copyright © 1984 by the President and Fellows of Harvard College.

"In Praise of Sharks," from an article of that title in *Discover,* 6 (July 1985): 40–53. Copyright © 1985 Discover Magazine.

"In the Company of Ants," *Bulletin of the American Academy of Arts and Sciences,* 45, no. 3 (1991): 13–23.

"Ants and Thirst," published as "Altruism and Ants," in *Discover,* 6 (August 1985): 46–51. Copyright © 1985 Discover Magazine.

"Altruism and Aggression," published as "Human Decency Is Animal," *New York Times Magazine,* October 12, 1975, pp. 38–50.

"Humanity Seen from a Distance," published as part of "Comparative Social Theory," in *The Tanner Lectures on Human Values,* vol. 1 (Salt Lake City: University of Utah Press, 1980), pp. 51–58. Reprinted courtesy of the University of

Utah Press, Cambridge University Press, and the Trustees of the Tanner Lectures on Human Values.

"Culture as a Biological Product," published as "The Biological Basis of Culture," in Joseph Lopreato, ed., *Sociobiology and Sociology,* a special monograph in *Revue internationale de sociologie,* n.s., 3 (1989): 35–60.

"The Bird of Paradise: The Hunter and the Poet, Science and the Humanities," from "The Bird of Paradise," in *Biophilia* (Cambridge, Mass.: Harvard University Press, 1984), pp. 51– 55. Copyright © 1984 by the President and Fellows of Harvard College.

"The Little Things That Run the World," published in *Conservation Biology,* 1 (1987): 344–346. Reprinted by permission of Blackwell Science, Inc.

"Systematics Ascending," published as "The Coming Pluralization of Biology and the Stewardship of Systematics," *BioScience,* 39 (1989): 242–245. Copyright © 1989 American Institute of Biological Sciences.

"Biophilia and the Environmental Ethic," published as "Biophilia and the Conservation Ethic," in S. R. Kellert and E. O. Wilson, eds., *The Biophilia Hypothesis* (Washington, D.C.: Island Press, 1993), pp. 31–41.

"Is Humanity Suicidal?" published in the *New York Times Magazine,* May 30, 1993, pp. 24–29.

INDEX

Aboriginals (Australia), 26
Adaptability of sharks, 42–43
Adaptive radiation, 35–38
Adoption, 78
Aesthetic value, 176
Africa, 22, 143, 192, 196
African gaboon vipers, 11
Aggression, 83–88
Agkistrodon (snakes), 11, 16–17
Alpha wave blockage, 118–19
Altmann, Stuart, 88
Altruism, 75–83
Amazon rain forest, 48, 142
Amino acid sequencing, 153
Amphibians, 144
Analytic/synthetic science, 133–35
Anatomy of Human Destructiveness (Fromm), 84
Androcentric culture, 191
Angel sharks, 37
Anthropocentric ethic, 176
Anthropologists, 100
Antiquity of the invertebrates, 142
Ants:
- aggressive acts, 86
- army, 38, 141–42
- biomass, 48, 142
- driver, 157
- food-storage caste, 67–68
- guidance systems, 69
- Gulf Coast, 10
- humans, teaching, 59
- leafcutter, 55–58, 143
- mites, 141–42
- neurobiology, 157
- social systems, 47, 49–52
- stinging, giant tropical, 63–65
- suicide, altruistic, 79–80
- weaver, 52–55

Aplysia californica (snails), 157
Arabian Sea, 43
Army ants, 38, 141–42
Art, role of, 129
Arthropods, 171
Asclepius (myth), 26
Ashtoreth (spirit), 26
Asia, 22
Assembly rules, 198

Asteroid colliding with Earth, 196
Atmosphere, 190, 193
Atta (ants), 55–58
Australia, 26, 38–39, 196
Australopithecus afarensis-Homo habilis (man), 104
Aztec people, 27

Baboons, hamadryas, 90
Bacteria, 154, 155–56, 171, 172, 177
Bees, 79–80
Beetles, 141, 153
Behavior, evolutionary origins of, 19–20
Berlin, Brent, 117
Bias in learning, 119
Biocultural evolution, 167–69
Biodiversity, 33, 35–38, 155, 170–79, 197–98
Biological channeling of culture, 116
Biology, 149–51
  evolution, 33
  neurobiology, 155–58
  pluralization of, 150–55, 160
  systematics, 158–61
  thematic shift, 149
Biomass, insect, 48, 142
Biophilia (Wilson), 7
Biophilic traits, 7–8
  environmental despoliation, 170–74
  environmental ethic, 175–79
  gene-culture coevolution, 167–69
  learning rules, 165–66
  testing thesis on, 169–70
Bird of paradise, 131–45
Birds, 35–36, 77, 101–2, 120, 141, 173
Black ants, 67
Black racers (snakes), 21
Black-tipped sharks, 37
Blindness, color, 117
Blue sharks, 37
Brain, the:
  ants, 68–69
  biocentric world, 166
  evolution, 29–30, 109
  juggernaut theory of human nature, 186
  light intensity, 116
  semantic memory, 113–14
Bramble sharks, 36
Brazil, 48, 142
Bull sharks, 38
Burma, 22
Bushmen, African, 87

Caduceus, 25–26
California, 39, 196
Camponotus (ants), 51
Cannibalism, 38, 86
Carcharodon carcharias (sharks), 38–42
Carnivores, 185

Cell biology, 149–51
Central America, 22–23, 143
Cerebral cortex, 109
Chemical secretions from ants, 54, 57
Chemoreception, 152
Children, snakes/serpents and young, 20
Chimpanzees, 20, 23, 77–78, 92–93, 166
Chinese, Han, 26
Chlorofluorocarbons (CFCs), 193
Cihuacoatl (goddess), 27
Civil War and disease, 10
Climate-controlled ant nests, 51
Coatlicue (god), 27
Coelenterates, 143
Cognitive development, 107, 115–19, 122–26
Colon bacteria, 154
Colonies, creating ant, 55–56
Colors, 114, 116–18
Concepts, 113, 114
Confederate army and disease (Civil War), 10
Congressional Medals of Honor, 75
Contraceptives, 192–93
Cookie-cutter sharks, 34–35
Copepods, 143
Coral reefs, 143, 172, 190, 196
Coral snakes, 11
Costa Rica, 192
Counterprepared bias in learning, 119
Crocodile sharks, 36
Crustaceans, 144
Cultural determinism, 89
Culture:
  altruistic acts, 82
  androcentric, 191
  as a biological product, 107
  epigenetic rules, 115–19
  genetics, coevolution with, 109, 114, 126, 167–69
  Lamarckian transmission, 101
  prescientific peoples, 25–26
  sharks, 44
  snakes/serpents, 12–13, 23, 28–29
  translation from genes to, 119–26
  units of, 111–15

Dangerous Reef (Australia), 39
Dani people (New Guinea), 118
Darwin, Charles, 149
Darwinism, see Natural selection
Death and birth schedules of leafcutter ants, 58
Decision-making/individual learning and cultural diversity, 122–26
Density-dependent control of rodent population, 152
Determinism, cultural/genetic, 89

Developmental growth and fear of snakes/serpents, 20–21
Dialectical synthesis, 149
Diamond, Jared, 169
Diamondback rattler, 13
Distributions, single/multiple mode, 125–26
Diversity of life, 33, 35–38, 155, 170–79, 197–98
Dogs, African wild, 77, 93
Dolphins, 77
Dorylus (ants), 157
Dorymymex (ants), 51
Dreams, 8–9, 24
Driver ants, 157
Drop-carrying in ant colonies, 65–67
Drosophila (flies), 156, 157
Dual-track system of human evolution, 109–11
Dwarf sharks, 35

Earth Day Summit (1992), 191–92
Echinoderms, 196
Ecofeminism, 191
Ecosystems, natural, 190
Eel-like frilled sharks, 37
Egyptians, ancient, 26
Eisner, Thomas, 67
Electron microscopy, 153
Emotions, 82, 114
Emperor of Germany bird of paradise, 131–35
Energy expenditure of leafcutter ants, 57
Engels, Friedrich, 149
Environment, stabilizing the global, 199
Environmental despoliation, 170–74, 181–86
Environmental ethic, 175–79
Environmentalism, 189–91, 198–99
Environment influencing aggression, 85, 88
Environment influencing culture, 121
Epigenetic rules, 110, 111, 115–19, 126
Erinyes (spirit), 28
Escherichia coli (bacterium), 155–56
Ethic, environmental, 175–79
Ethics, naturalistic fallacy of, 93
Ethnographic distribution, 124–25
Eukaryotic organisms, 177
Euripides, 28
Europe, 23
Evolution:
  behavior, origins of, 19–20
  biology, 33
  brain, the, 29–30, 109
  evolutionary grades, 154
  gene-culture coevolution, 109, 114, 126, 167–69
  logic, 166, 176

Exemptionalism, 188–89, 197–98
Exotic species, 172, 194
Exploration drives, 160
Extended relationships in hunter-gatherer societies, 91

Facial expressions, 114
Farallon Islands (California), 39
Farancia (snakes), 11
Feeling and myth, 134–35
Female ant societies, 53
Fertilization in sharks, 42
Finches, Galápagos, 36
Finland, 23, 52
Fish, 144, 194
Florida, 10–12
Food-storage caste in ant colonies, 67–68
Food/water sharing, 63–68, 77
Forests, 48, 130–32, 142, 144, 171, 196
Formica subsericea (ants), 67
Frequency-distribution function and human behavior, 100–101
Freudian theory, 24–25
Fromm, Eric, 84
Fu-Hsi (demon), 26
Fungi, 55–58, 144, 171
Future, biodiversity as frontier of the, 178–79

Gaia, 144, 190
Galápagos Islands, 36
Galeocerdo cuvieri (sharks), 34
Garden of Eden, 28
Gender and human patterns, 91, 92–93
Genetics:
ants' social system, 52
bacteria, 172
biological controlling devices, 29
conventional transmission, 101
culture, coevolution with, 109, 114, 126, 167–69
culture and translation process, 119–26
determinism, 89
emotions, 82
juggernaut theory of human nature, 186
mapping, 154
natural selection, 107–8
neurobiology, 156
social theory, 102–4
sociobiology, 76
Geographic variation across human populations, 103
Geometric forms, 114, 118–19
Gill slits, sharks, 42
Global warming, 193
Globitermes sulfureus (termites), 80
Goodall, Jane, 78
Grasslands of Africa, 143
Great white sharks, 38–42

Greek myths, 26, 27–28
Greenhouse gases, 193
Guidance systems in ant colonies, 69
Gulf Coast, 10–12
Gulper sharks, 36
Gynandromorphs, 156

Habitat alteration/destruction, 173, 194
Hammerhead sharks, 38
Harvard's Museum of Comparative Zoology, 48
Hawaii, 36, 194
Hearing, 110
Herbivores, 143, 185
Hereditary basis of social behavior, 76
Heredity, nature of, 89–94
Herpetology, 150, 151
Heuristic value, 152
Hindu people, 26
Hognose snakes, 11, 12–13
Hölldobler, Bert, 52, 65, 66–67
Holometabolous insects, 152
Holton, Gerald, 149
Homo habilis (man), 64
Homo sapiens (man), 184
Homosexuality, 82–83
Honeybees, 79–80
Honeycreepers, Hawaiian, 36
Hoop snakes, 13
Hopi people, 8, 88
Humanities, 99–100, 134
Human nature, 29–30, 82, 186
Humans:
  aggression, 85
  ants teaching, 59
  environmentalism, 189–91, 198–99
  exemptionalism, 188–89, 191–98
  gender and human patterns, 91, 92–93
  geographic variation across human populations, 103
  hazards, environmental, 186
  homosexuality, 82–83
  hunter-gatherers, 83, 90–92, 166, 186
  kinship between animals and, 177
  meat eating, 185
  nonliving physical environment, 193
  optimistic trends coming from, 191–93
  population issues, 184, 187
  reassembling biodiversity, 197–98
  scientific knowledge, 188
  self-image as a species, 191
  snakes/serpents, innate fear of, 6
  social repertoire, 101–2
  species, reduction in number of animal, 173, 194–96

Hunter-gatherers, 83, 90–92, 166, 186
Huon Peninsula (New Guinea), 129–35
Hyenas, 85–86
Hymenopterous insects, 47

Illugason, G. S., 43
Immunochemistry, 152
Inanimate matter *vs.* life, 7
India, 26
Indonesia, 192
Inductive descriptions of behavior and culture, 122
Infanticide, 154
Insects, social, 47, 49–50, 157
see also Ants
Intellectual drives, 160
Intimate group members, 91
Invertebrates, 141–45, 171, 173
Iphigeneia in Tauris (Euripides), 28
Ireland, 23
Isistius brasiliensis (sharks), 34–35

Judaism, 28
Juggernaut theory of human nature, 186

Kay, Paul, 117
Kingsnakes, 11, 38
Kin selection, 81–83, 152
Kinship between animals and humans, 177
Krogh's rule, 154–55
Kruuk, Hans, 85–86
K strategy of reproduction, 108–9
Kwakiutl people, 8, 26

Labor, leafcutter ants' division of, 56–57
Labor, women/men and division of, 91, 92–93
Lamarckian transmission, 101
Langurs (monkeys), 85
Larvae, ant, 55
Leafcutter ants, 55–58, 143
Learning, bias in, 119
Learning rules and biophilic traits, 165–66
Lemurs, 20
Level-oriented disciplines, 150–53
Light intensity, 116–17
Lily pond riddle, 188
Liodytes (snakes), 11
Lions, 85, 93
Livers, shark, 42–43
Logan, Frank, 41
Lorenz, Konrad, 84
Lumsden, Charles, 122, 125, 167

Madagascar, 20
Magic spells, 25–26
Malaria, 10
Malaysia, 194
Mammals, 141, 143, 144, 171, 173
Manasā (goddess), 26

Mandarin dogfish (sharks), 36
Marine invertebrates, 171
Markov processes, 123
Marriage ceremonies, 115
Maternal care in hunter-gatherer societies, 91
Mathematical ability, 92
McCosker, John, 39–40, 42
Meat eating, 185
Mediterranean-type heathland, 172, 196
Megachasma pelagios (sharks), 44
Megafaunas, 173
Megamouth sharks, 44
Meilichios (god), 27
Memory, semantic, 112–15
Mendelian genetics of color blindness, 117
Mental development, 22–25, 113, 121
Mercury (myth), 25–26
Mesozoic era, 130
Metaphor, 166
Metaphysical constructs, 149
Microniches, 141–42
Mites, 141–42
Molecular biology, 149–51
Mollusks, 144, 196
Monkeys:
  aggression, 85, 87–88
  altruism, 77–78
  biophilic traits, 166, 167–68
  hunter-gatherers' qualities compared to, 90–91
  labor, gender and division of, 92–93
  snakes/serpents, responses to, 18–20, 23
Moral values, 99, 175
Mosquitoes, 10–11
Mound-building ants, 52
Mudammā (goddess), 26
Multidimensionality, 100–101
Multiple mode distributions, 125–26
Multivariate analysis, 153
Mundkur, Balaji, 26, 167
Mussels, 194
Mutation pressure, 108
Mutations, cultural/genetic, 121
Myrmecocystus (ants), 51
Mythology, 25–26, 134–35, 166, 168

Natrix (snakes), 11, 14–15
Natural heritage, biodiversity as a country's, 177–78
Naturalistic fallacy of ethics, 93
Natural kinds/units, 111–15
Natural selection:
  biophilic traits, 167
  conventional genetic transmission, 101
  genetic evolution driven by, 107–8
  homosexuals, 83
  Huon forest, 132
  self-sacrifice, 81

snakes/serpents, aversions to, 19–20
Nehebkau (god), 26
Nematology, 150, 151
Nephila (spiders), 10
Nests, ant, 51–52
Neurobiology, 155–58
New Guinea, 21–22, 129–35
Newton, Isaac, 149
Nigeria, 187
Nonliving physical environment, 193
Nuclear warfare, 93
Nu-kua (demon), 26
Nuptial flights, ant, 55–56

Oceans, 190
On Aggression (Lorenz), 84
Ophidian archetype, 9
Ophidiophobia, 21
Orthogenesis (straight-line evolution), 108
Ostyak people (Siberia), 26
Otters, 42
Ozone layer, 193

Pachycondyla (ants), 63–65
Paleohominid times, 166
Paradisaea guilielmi (birds), 131–45
Pavement ant, 86
Pavilions, ant, 53, 54–55
Personality, 107
Pheidole (ants), 48
Phenotypic variation, 114
Phobias, 119, 168
Phylogenetic reconstruction, 160
Phylogeny, 154
Physiology, 29
Piagetian stages of mental development, 113
Pit vipers (snakes), 22–23
Plankton, 35, 143
Plants, 144, 173, 194, 196
Play/contest/mock-aggression aspect of normal development, 91
Plovers, golden, 101–2
Pluralization of biology, 150–55, 160
Poisonous snakes, 22
Pollution, 172
Population issues, 93, 184, 187, 190, 192–93, 199
Population level of organization, 153
Possibilities/probabilities, inheriting a pattern of, 90–94
Poverty, 193
Precambrian times, 142
Prepared bias in learning, 119
Prescientific peoples, 25–26
Primatologists, 100
Private space, 184
Probabilistic operation of the mind, 124
Promethean Fire (Lumsden & Wilson), 122

Propositions, 113, 114
Psychic unity of humankind, 102
Psychoaesthetics, 118–19
Psychoanalysis, 25
Pure cultural/genetic transmission, 120
Pygmy rattlesnakes, 15–16
Pygmy sharks, 36

Queens in the ant colony, 50, 51, 53, 55–56, 58, 157
Quetzalcoatl (god), 27

Racial differences in motor/temperament development of newborns, 103
Rain forest, 48, 142, 171, 196
Rattlesnakes, 11–13, 15–16
Reassembling biodiversity, 197–98
Recruitment systems of weaver ants, 54
Regurgitating food, 66–68, 77
Religion, greening of, 192
Religion and snakes/serpents, 12, 26–29, 168
Reproductive strategies, 108–9, 184–85
Reptiles, 141
Resources, 188, 199
Retina, 116–17
Rhesus monkeys, 87
Rhincodon typus (sharks), 35
Ribbon snakes, 11
Rights of species, 175–76
River systems, 172
Rodents, 152
Rosch, Eleanor, 114, 117
R strategy of reproduction, 108–9

Scavengers, ant, 49
Schemata, 113, 114
Science, 129, 133–35, 188, 191
Sea lions, 39, 40
Seals, 39, 40, 42
Self-sacrifice, altruistic, 75–83
Selket (goddess), 26
Semantic memory, 112–15
Seminatrix (snakes), 11
Serpents, *see* Snakes/serpents
Sexual drives, 184–85
Sharanahua people (Peru), 8–9
Sharing food/water in ant colonies, 63–68
Sharks:
- adaptation skills, 42–43
- adaptive radiation, 35–38
- eating habits, 39–41
- great white, 38–42
- ignorance about, 43–44
- new species, 44
- species types, 33–34

Siberia, 26
Silk-spinning ants, 55
Single mode distributions, 125–26
Sleeper sharks, 36
Smell, sense of, 53–54, 57, 110

Smets, Gerda, 118
Snails, 157, 194
Snakes/serpents:
  biophilic traits, 167–69
  cultural lore around, 12–13, 23
  dreams, 8–9
  eating members of own group, 38
  Gulf Coast, 10–12
  handling routines, 17
  human nature, 29–30
  innate fear of, 6, 18
  mental development, 22–25
  monkey's responses to, 18–20
  prey eaten, 11
  religion, 26–29
  science and humanities bridged by, 5
  watersnakes, 11, 14–17
Social insects, 47, 49–50, 157
  see also Ants
Social sciences, 99–100
Social theory, 102–4
Sociobiology, 76, 88–89, 90, 93, 94
Sociologists, 100
Soil, 144, 145
Solitary insects, 50
South America, 22–23, 143
Space scientists, 190
Sparrows, white-crowned, 120
Specialists living in microniches, 141–42
Species, demographics on, 153–54
Species, reduction in number of, 172–74, 194–96
Species-specific nature and morality, 99
Sphecomyrma (ants), 49
Spiders, 10
Spiritual value, 176
Stabilizing the global environment, 199
Stingless bees, 79–80
Stylized snakes, 26
Suicide, altruistic, 79–80
Sun's energy, 185
Superorganisms, 157–58
Switzerland, 23
Symbiosis among the disciplines, 153
Synthetic science, 133–35
Systematics, 158–61

Taste, sense of, 53–54, 110
Taxonomic groups of organisms, 149, 150, 159–61
Technology, 188, 191
Teeth, shark, 42
Termites, 48, 80, 97–99, 142, 175
Territorial behavior, 91
Tetramorium caespitum (ants), 86
Themata of science, 149
Thresher sharks, 37
Tiger sharks, 34, 38
Tiger snakes, 22
Tlaloc (god), 27

Tribal security, 184, 186
Troy, Joseph, Jr., 41

Ubiquity of snakes, 22–23
Ulrich, Robert, 169
Understanding of life as a whole, 153–54
United Nations, 43, 191–92
Utilitarian potential of wild species, 176

Vegetation, 144
Verbal ability, 92
Vertebrates, 141–44
Vipers (snakes), 11, 22–23
Viperus berus (snakes), 23
Vision, 110, 116–18

Warlike behavior of ants, 50–51
Warming, global, 193
Warning to Humanity, 192
Wasp ants, 49
Watersnakes, 11, 14–17
Weaver ants, 52–55
Whale sharks, 35, 43
Wheeler, William M., 79
Whitehead, Alfred N., 160
White sharks, great, 38–42
Wobbegongs (sharks), 36, 37–38
Wolves, 93

Xiuhcoatl (god), 27

Yakut people (Siberia), 26
Yellow fever, 10
You Shall Know Them (Vercors), 97

Zeus, 27
Zimbabwe, 192
Zoologists, 100
Zoos, 165, 168